John Wrightson

The principles of agricultural practice as an instructional subject

John Wrightson

The principles of agricultural practice as an instructional subject

ISBN/EAN: 9783337124045

Printed in Europe, USA, Canada, Australia, Japan

Cover: Foto ©Lupo / pixelio.de

More available books at **www.hansebooks.com**

THE PRINCIPLES

OF

AGRICULTURAL PRACTICE

AS AN

INSTRUCTIONAL SUBJECT.

BY

JOHN WRIGHTSON, M.R.A.C., F.C.S., &c.,

EXAMINER IN AGRICULTURE TO THE SCIENCE AND ART DEPARTMENT; PROFESSOR OF
AGRICULTURE IN THE NORMAL SCHOOL OF SCIENCE AND ROYAL SCHOOL OF MINES;
PRESIDENT OF THE COLLEGE OF AGRICULTURE, DOWNTON, NEAR SALISBURY;
LATE COMMISSIONER FOR THE ROYAL AGRICULTURAL SOCIETY
OF ENGLAND, ETC., ETC.

Third Edition.

LONDON: CHAPMAN AND HALL, LD.

1893.

[All rights reserved.]

PREFACE.

THE province of the teacher is to communicate elementary as well as advanced knowledge. Whatever he may do in the form of original and elaborate research, to his pupils he must ever appear as a patient expounder of first principles. The following chapters actually embody a course of lectures delivered to science teachers last summer. The intention was to open up the subject of agriculture in such a manner as to indicate to them the many aspects from which it may be viewed. It has always been difficult to accurately define the position which *agriculture* occupies in relation to other branches of knowledge. Sometimes it has been called a science, and at other times an art. It has been viewed as a trade and as a profession, but it is neither one nor the other, and in order to be correct we must take refuge in such ambiguous expressions as occupation, pursuit, or calling. Agriculture is too dependent upon circumstances, and too uncertain in its results, to be viewed as a science, and too natural to be called an art. It is not, strictly speaking, a trade, and is not included among the professions. It is more ancient than knowledge, and lies fundamentally at the basis of our very existence. In addition to hunting as a means of sustenance, some rough-and-ready control of vegetation appears as the next step in the direction of regular supplies of food, and Agriculture emerges from the shadow of the forest, or descends from the mountain slopes to take possession of the plain. Cain was a tiller of the ground before Tubal Cain became an instructor of artificers in brass and iron, and the two branches of grazing and tillage were twin-born.

Agriculture seems to have always been the recipient of benefits from both art and science, but has preserved her individuality throughout. She has hoarded up maxims, perfected processes, improved her products, and adapted herself to all descriptions of soil and climate. She has subjugated animals, and modified their forms, habits, and aptitudes. She has invented implements, contrived rotations, and discovered fertilizers. She is the mother of Horticulture. Those who would study the principles of Agriculture must learn her maxims, become familiar with her processes and products, and master her variations; they must be familiar with her implements, courses of cropping, fertilizers, and methods; they must to some extent be able to exercise sound judgment upon her ways, and offer their share of suggestion. It is this broad knowledge of agriculture which it is my object to inculcate. I have endeavoured to show that agriculture is in itself a large and varied subject, not to be mastered in the laboratory or the lecture-room, but rather in the field and in the fold; that the Professor of agriculture must be a Professor of farming as well as an adept at science; and that he must never fail in his allegiance, or forget that he is retained in the service of improved cultivation. Let others throw side-lights upon his subject, and let him thankfully accept the helpful ray; but let him beware of being dazzled by the light, and of removing his eye from where it falls, to wander in search of its source. This is the grand evil of the application of science to agriculture, and it has proved a pit-fall both for those who profess and for those who study her. A theorist or a scientific agriculturalist are too often synonyms for those who are ignorant of agriculture as a pursuit, and who pose rather as scientists explaining phenomena, than as enlightened agriculturalists teaching old and new methods.

The following pages only deal with a small and fundamental portion of the vast subject of agriculture. So far as they go, they are intended to point out the principles which guide the practice of agriculturalists and the considerations which weigh with them. They are intended also to gratefully acknowledge and appreciatively notice the bearings of science upon our pursuit. A chemist cannot profess agriculture, and an agriculturalist cannot profess chemistry; but the interdependencies as well as the individuality of each subject must be acknowledged and respected.

I must ask my readers' indulgence for the method of treatment, which is somewhat different from what would have been adopted had these pages emanated entirely from the pen. They were delivered orally, and are published by request. They are consequently popular in style, and a certain amount of summary and repetition became necessary.

The chapters are indeed lectures, and each chapter opens with a reference to what had previously occupied attention. Omissions of certain principles made during the delivery of one lecture were often supplied upon the next occasion, whereas in the ordinary course of writing, an exhaustive treatment would have been preferred. It is my hope to follow up the present volume with another, and to open up, to the best of my ability, the proper lines upon which the principles of the whole art of Agriculture should be taught.

JOHN WRIGHTSON.

Downton, April, 1888.

CONTENTS.

CHAPTER I.

NARROW AND INCOMPREHENSIVE METHODS OF TEACHING AGRICULTURE—BEARINGS OF VARIOUS BRANCHES OF SCIENCE ON AGRICULTURE—THE BOUNDARIES AND LIMITS OF AGRICULTURE PROPER—ITS HISTORY—" COMPARATIVE AGRICULTURE "—PROCESSES OF AGRICULTURE—RULES AND AXIOMS OF AGRICULTURISTS—STATISTICAL ASPECT—ROTATIONS OF CROPS—DESIGNING OF BUILDINGS—TILLAGE—IMPLEMENTS—LABOUR—COSTS—DAIRYING—BREEDING AND MANAGEMENT OF ANIMALS—ORIGIN AND DISTRIBUTION OF SOILS—FERTILITY—MANURING—MISLEADING TEACHING—THE BEST GUIDES FOR TEACHERS—JOURNALS OF THE SOCIETIES—SYLLABUS OF THE SCIENCE AND ART DEPARTMENT ... 1

CHAPTER II.

AGRICULTURE AS A SUBJECT—DEFECTIVE TREATMENT OF IT—PROPER FUNCTION OF THE AGRICULTURAL TEACHER—STUDY OF THE SOIL 17

CHAPTER III.

STUDY OF THE SOIL (*continued*)—ITS COMPLICATED CHARACTER—SOLUBLE MATERIAL—FINE IMPALPABLE MATERIAL—QUARTZOZE MATERIAL—CALCAREOUS MATERIAL—VEGETABLE MATTER—STONES—COMPLEX FUNCTIONS OF SOILS—CONDITIONS OF FERTILITY—INDICATIONS OF FERTILITY ... 26

CONTENTS.

CHAPTER IV.

PAGE

INDICATIONS OF FERTILITY (*continued*)—THE SUBSOIL—GEOLOGICAL POSITION OF SOILS AFFECTING FERTILITY—GEOLOGICAL KNOWLEDGE TO BE USED WITH CAUTION—GEOLOGICALLY RECENT SOILS 49

CHAPTER V.

SOILS OF THE LONDON CLAY—SOILS OF THE CHALK FORMATION 65

CHAPTER VI.

GEOLOGICAL SECTION FROM HERTFORD TO BRIDGWATER—SOILS OF THE GREEN SAND FORMATIONS—THE GAULT CLAY—THE WEALD—THE UPPER OOLITE—THE MIDDLE OOLITE—THE LOWER OOLITE—THE LIAS—THE NEW RED SANDSTONE—THE PERMIAN FORMATION—THE COAL MEASURES—THE MOUNTAIN LIMESTONE—THE MILLSTONE GRIT AND YOREDALE ROCKS—THE OLD RED SANDSTONE 71

CHAPTER VII.

FORMATIONS YIELDING SOILS OF CLAYEY, FREE, POOR, AND RICH CHARACTER—REVIEW OF INDICATIONS OF FERTILITY—DIFFERENCES BETWEEN QUALITY AND VALUE OF LAND—SEDENTARY AND TRANSPORTED SOILS—PEAT SOILS—VOLCANIC SOILS—METHODS OF IMPROVING SOILS—LAND DRAINAGE—REASONS FOR ITS USEFULNESS 87

CHAPTER VIII.

BENEFICIAL EFFECTS OF VEGETATION ON SOILS AFTER DRAINAGE—ACTION OF DRAINS IN LIGHT SOILS—ACTION OF DRAINS ON RETENTIVE SOILS—SMITH'S SYSTEM—ELKINGTON'S SYSTEM—ADVANTAGES OF DRAINAGE—IMPROVEMENT OF LAND BY TRENCH PLOUGHING—SUBSOILING—CLAY-BURNING—CLAYING—MARLING—CHALKING—WARPING—ORDINARY CULTIVATION 97

CONTENTS. xi

CHAPTER IX.

ORDINARY CULTIVATION (*continued*)—MAIN DIFFERENCES BETWEEN THE CULTIVATION OF STIFF AND LIGHT SOILS—ROOT CULTIVATION—AUTUMN CULTIVATION—HOW TO CLEAN A FOUL FIELD 118

CHAPTER X.

CLASSIFICATION OF CROPS—PRINCIPLES OF CULTIVATION FOR ROOT CROPS—FOR CORN CROPS—FOR GRASS CROPS—FOR "FODDER" CROPS AND "CATCH" CROPS—SYLLABUS OF CROP CULTIVATION 132

CHAPTER XI.

FERTILIZERS—RELATION BETWEEN MECHANICAL AND CHEMICAL METHODS OF LAND IMPROVEMENT—LIEBIG'S VIEWS—GENERAL MANURES—FARMYARD MANURE—M. VILLES' VIEWS—ROTHAMSTED RESULTS—WHY FARMYARD MANURE IS ESTEEMED—ITS CHEAPNESS—ARTIFICIAL MANURES—DEFECTIVE TEACHING AS TO MANURING—EFFECTS DEPENDENT UPON CONDITION OF SOIL—AND OF CLIMATE ... 146

CHAPTER XII.

ACTION AND RE-ACTION OF FARMYARD DUNG THROUGHOUT A ROTATION—DISADVANTAGES FROM USE OF GENERAL MANURES—OBJECTIONS TO USE OF SPECIAL MANURES—PERMANENT EFFECT OF ONE CROP MANURES—PRINCIPLES IN APPLYING SPECIAL MANURES—ALLEGORICAL EXPRESSIONS TO BE AVOIDED—EFFECTS OF PHOSPHATIC AND OTHER MINERAL MANURES UPON THE PRINCIPAL FARM CROPS—SIMILAR EFFECTS OF NITROGENOUS MANURES—ROTHAMSTED RESULTS—EFFECTS OF FERTILIZER ON GRASS LAND—BASIC CINDER—PRINCIPLES IN CONDUCTING FIELD EXPERIMENTS ... 160

CONTENTS.

CHAPTER XIII.

PAGE

ROTATIONS OF CROPS—PRIMITIVE ROTATIONS—DEVELOPMENT OF ROTATIONS—PRINCIPLES OF ROTATIONS—THE FALLOW—GRAIN AND FODDER CROPS—EFFECT OF CLOVER ON SUBSEQUENT WHEAT CROPS—SIMILAR EFFECT OF BEANS—MODIFICATIONS OF THE NORFOLK FOUR COURSE—CATCH CROPS—POTATOES AS A CROP 181

CHAPTER XIV.

DEPRECIATED VALUE OF CLAY LANDS—HOW TO MAINTAIN LIVE STOCK ON CLAY SOILS—CATCH CROPPING A MATTER OF SITUATION—SOIL AND CLIMATE—THE THEORY OF ROTATIONS—CLOVER SICKNESS—PRACTICAL ADVANTAGES OF ROTATIONS—PURIFYING EFFECT ON LAND FOR SHEEP—ADVANTAGES OF LIGHT SOIL—LAYING LAND DOWN TO GRASS—ITS DIFFICULTIES—ITS EXPENSE—ITS TEDIOUSNESS—HOW BEST TO BRIDGE OVER THESE DIFFICULTIES ... 200

CHAPTER XV.

PROPER STANDARD FOR VALUING TILLAGE OPERATIONS—THE COST OF MAINTAINING A FARM HORSE—THE COST OF TILLAGES—THE COST OF GROWING A CROP OF WHEAT AFTER CLOVER—SMALL MARGIN OF PROFIT FROM WHEAT GROWING ... 216

THE PRINCIPLES OF
ENGLISH AGRICULTURE.

CHAPTER I.

Narrow and Incomprehensive Methods of Teaching Agriculture—Bearings of Various Branches of Science on Agriculture—The Boundaries and Limits of Agriculture Proper—Its History—"Comparative Agriculture"—Processes of Agriculture—Rules and Axioms of Agriculturists—Statistical Aspect—Rotations of Crops—Designing of Buildings—Tillage—Implements—Labour—Costs—Dairying—Breeding and Management of Animals—Origin and Distribution of Soils—Fertility—Manuring—Misleading Teaching—The Best Guides for Teachers—Journals of the Societies—Syllabus of the Science and Art Department.

THE object of this short treatise is to show that the principles of agriculture are capable of being taught in the lecture-hall or the school-room, fully bearing in mind that agriculture is an occupation requiring experience for its successful prosecution. A disposition has been exhibited on the part of those who have undertaken the explanation of matters agricultural to treat them in a somewhat narrow and incomprehensive manner. Sufficient prominence has not been allowed to the fact that agriculture is itself a subject. The various natural sciences shed a glorious light upon the practice of agriculture, but we must be careful to discriminate between them and Agriculture herself. There is a danger of presenting the subject as though it were a patchwork or

B

'mosaic composed of fragments of all the known sciences. But while the agricultural teacher ought to be well instructed in the various sciences bearing upon agriculture, he ought not to forget that he has chiefly to do with a great central subject upon which modern science throws its beneficent rays. The necessity of such a warning has been forced upon me chiefly in my capacity of Examiner in Agriculture to the Science and Art Department. I have been struck with the degree of narrowness, not only in the answers given to questions, but also in the selection of questions to be answered. It may be well to explain that the Department allows a considerable amount of choice to candidates for its honours. The examiners may set twelve, or even fifteen, questions, but the regulations of the Department only allow the candidate to answer a limited number, usually about six. It is natural that examiners should vary their questions so as to elicit information upon as large a number of topics as possible, but the bias of candidates is distinctly towards a certain class of questions dealing rather with the molecular, microscopic, and minute sections of the subject than with those of larger, wider, and more practical scope. I have been led to think how far the class of topics evidently preferred by candidates would be of interest to even highly-educated men who are intrusted with the ownership or, it may be, the management of large landed estates. From long contact with such men, I am inclined to think that a large number of the topics upon which examinees love to dilate would excite among them but a feeble interest. I am aware that those who present themselves as candidates for the honours of the Science and Art Department must have been influenced in their choice of questions by the teaching and the text-books to which they have had access, and the inference forced upon me is that their selection of questions, and their treatment of such questions, is a reflection of the instruction and the reading provided by their numerous teachers. I have, therefore, come to the conclusion that the teaching of agriculture has fallen into a groove, and that an effort should be made to place this large subject upon a wider basis. The Royal Agricultural Society of England may unconsciously have assisted in producing the pernicious result complained of by the fact that in its syllabus of agricultural education it places the word "chemistry" in a parenthesis immediately after the phrase "theory of agriculture," thereby conveying the idea that the theory of agriculture is chemistry. It will, however, be one of my first endeavours to show that chemistry is far from representing the theory of agriculture. Agriculture is not chemistry any more than chemistry is agriculture. If agriculture was to be defined as belonging to any particular domain or branch of science, it probably would be ranged more properly as a statistical subject than in any other position. Chemistry deals with compositions, and there is abundant scope for chemical investigation with regard to agriculture, whether considered in its bearings upon the composition of soils, the ingredients of the ash and of the soft tissues of plants, or the nutrient value of various seeds and feeding stuffs, the composition of the increase of fatting animals, or the genuineness of substances offered as fertilizers; in all of these we see the immense value of chemistry and of chemical knowledge. Again, all questions connected with the nutrition of plants and animals, and the sources of nitrogen, and the processes by which crude mineral and organic matter are converted into available plant food, are eminently chemical questions. But in the same manner it would not be difficult to show the vast importance of other sciences as well as of chemistry in relation to agriculture. What, for example, can be more important than the bearings of botany upon the pursuit of agriculture, the ramification of the roots of plants, the fertilization of seed, the detection of adulteration in mixtures of grass-seeds, the knowledge of plants which may be introduced into our agriculture, and the identification of injurious plants, either in the form of pernicious weeds, or as impurities in mixtures of seeds presented for sale by unscrupulous seedsmen, or it may be in the tracing of the life history of those cryptogamic or fungoid forms, which so often are the cause of loss to farmers—as, for example, the potato disease and other blights and mildews which affect our crops. This brings us naturally to the domain of vegetable nosology, a most important branch of botany, so that whether we view this great subject in relation to the structure of plants, the classification of plants, or the diseases of plants, we see its bearings upon the pursuit of agriculture. Scarcely less important is the subject of physiology, dealing as it does with the functions of animal and vegetable life. So also it would be difficult to convey anything like an adequate idea as to the history and attributes of soils without importing a good deal of knowledge which belongs essentially to the science of geology. It is not my object to pass in review every natural science which throws light upon agricultural problems or agricultural practice; it is sufficient to show that there are many branches of natural science which cannot be overlooked by any intelligent student of agriculture. But while the natural sciences are of vast importance in elucidating difficulties and suggesting improvements, it can scarcely be denied that the great subject of agriculture requires for its full explanation a knowledge of mathematics, of political economy, of statistics, of engineering, and of physics, and there is perhaps scarcely a branch of knowledge which might not be easily shown to have a practical and important bearing upon the usages and customs of agriculture. Take, for example, meteorology, and even astronomy. Consider for one moment the influence of the sun as a factor in the germination of seed, and then reflect upon the nature of sun-heat and sun-light, and we shall find that plant and animal life are intimately connected with the phenomena essentially belonging to the domain of the physicist, the astronomer, and the electrician. And yet it would be most unfortunate that the agricultural teacher should find himself in the position of attempting to teach all of the above-named subjects, and probably of many others which have not been named, in his efforts to impart a sound knowledge upon the theory or principles of agriculture. He must confine himself to his particular province, and in order to do so successfully it would be well that he should endeavour to find a parallel by which to guide his own method. Take, for example, surgery, a subject upon which many sciences shed their light, and yet a teacher of surgery, while recognizing the importance of these branches, and while pointing out their bearings upon his own particular topic, must not fly off at a tangent and become a teacher of sanitary science, of pathology, or of therapeutics. A teacher of agriculture must keep his own subject steadily before him as a large practical occupation involving the expenditure of large sums of money, and a knowledge of mankind as well as of animals, vegetables, and the soil. The fact is that in order to teach the theory of agriculture a Faculty is required just as much as in the teaching of medicine, of law, or of divinity. It is too much to expect one man to grasp the whole of this worldwide subject, and therefore in any agricultural college or seminary it is necessary that chairs should be established for the teaching of chemistry, geology, physiology, anatomy, biology, physics, mathematics, estate management, building, and agriculture. The agricultural teacher ought to indicate the points of contact between his own subject and the sciences which surround it, rather than to pursue the explanation into those other domains.

If I were to endeavour to describe the boundaries and

limits of agriculture as a subject, I should in the first place begin with its history. The history of agriculture is a record of progress, and nothing could be a better incentive towards further advancement than the appreciation of the fact that much of what has been achieved has been done through intelligence and a true spirit of observation. The changes which have been wrought during the past two centuries are well worthy of attention, and the fact that the producing powers of the soil have been quadrupled through the efforts of agriculturists, that new and improved races of domesticated animals have been produced, that almost all of the crops which are cultivated in this country have been introduced from abroad, and that within comparatively recent times, the improvement in appliances of all sorts, the introduction of steam-power, and the improvement of new varieties of cultivated plants, are all important points in agricultural history, many of which are associated with the names of pioneers in agricultural progress. The history of agriculture is well calculated to stimulate to further improvement, and with this end in view ought not to be lost sight of, but on the contrary ought to be systematically taught if agriculture is a subject worthy of the teacher's attention. Another topic upon which much might be said, but which appears to be neglected by those who undertake to teach agriculture, may be spoken of as comparative agriculture, or the comparison of the practices of different districts, or even of different countries. The peculiarities noticeable in the management of land and of live stock in the North, the South, the East, and the West of England are worthy of attention; they are sometimes due to climate, sometimes to soil, and sometimes can only be attributed to custom. But these varieties of practice are exceedingly instructive, and their study might well lead to the introduction of improved methods, or the modification of old ones.

The processes of agriculture are many of them reducible

to description, and it scarcely seems possible to give students a just idea as to the theory of agriculture without giving them an intelligent idea as to the processes employed. Take for example the processes of the dairy, and the variations in practice in such a matter as that of cheese-making. Instead of confining himself to such chemical matters as the changes which take place when rennet is added to milk, the size of the fatty globules of milk, the serous fluids in which the globules of casein, or of butter fat, are suspended, it would seem necessary to go a step further and explain the processes of cheese-making as practised in the districts which yield us such excellent products as the Stilton, Cheshire, Gloucester, or Cheddar cheeses, and, if necessary, to go further afield and show the differences of manipulation requisite to produce the soft cheeses of the Continent, such as Camembert and Brie. By such a system of teaching he would lift his subject out of the narrow groove into which it appears to have fallen.

The process of the fattening of animals might similarly be enlarged upon with great advantage without confining the teaching to questions connected with the merits of nitrogenous foods as contrasted with farinaceous foods, and without considering the whole subject to be summed up in a treatise upon flesh-formers, and fat-formers or heat-producers. The dieting of farm stock, the times for feeding, the methods of accommodation, the cost, and the return, are all of interest, and if united with some instruction upon the nutrient properties of various food constituents would tend to give the subject a more practical aspect.

Much might be said upon the advantage of giving sound instruction upon the multifarious practices of the field and the homestead, from the treatment of land in order to produce the essential conditions for successful germination onwards to the cultivation and after cultivation, the securing or harvesting, and the preparation of crops for market, or it may be for home consumption. Similarly the treatment of dairy stock and of sheep, as well as of other animals, are all underlaid by principles or rules which are accepted by all good agriculturists. These principles and these rules which guide us in the conduct of successful farming business are capable of enunciation, and fall properly within the domain of the Professor of Agriculture, and as certainly fall outside that of the chemist or the biologist.

The statistical and economical aspects of agriculture ought not to be neglected, and, as already stated, agriculture frequently appears rather in the light of a statistical subject than of any other.

The subject of rotations of crops is one which might well occupy an agricultural class for weeks. Questions upon rotations are as a rule either shirked or treated without the least judgment. There is no appreciation of the fact that our various crops are suited to different conditions of soil and climate. It is very usual to see "roots" put down as the commencing crop of a series either for light or for heavy land, the candidate apparently having no idea that under the term "roots" a large number of crops may be ranged, and that such leafy crops as rape, cabbage, kale, kohl, may be introduced as helps or as substitutes to such crops in particular conditions of soil.

The knowledge of a candidate is shown by the kind of crop he suggests as suitable for a clay, a sand, a peat, or a limestone soil, and to get rid of the subject by using the term "roots" is as disappointing as if he were to use the expression "corn" without specifying whether he meant wheat, barley, oats, or rye.

The designing, and the considerations which should weigh in the designing, of farm buildings for the accommodation of our animals falls quite as much within the domain of agricultural teaching and agricultural theory as does the minute composition of the soil, or of the ash of plants. A knowledge of implements or of farm machinery would certainly be insisted upon by any examiner called in to test the amount of agricultural knowledge possessed by a candidate for agricultural honours. No person who is unacquainted with the peculiar merits of ploughing, cultivating, harrowing, or rolling, or with the various descriptions of instruments used in these processes, could be looked upon as having an adequate knowledge of the principles of agricultural practices. The instruments employed upon the farm are very numerous; they are required in the first instance for the preparation of the ground, in the second place for the drilling or distribution of seed, in the third place for the hoeing, or after cultivation, or after treatment of growing crops, subsequently for the harvesting or securing of those crops, and after that for the threshing, dressing, grinding, or other processes by which those crops are got up for sale, or are utilized for the feeding of live stock.

Questions connected with labour are capable of being discussed and taught. They divide themselves naturally into payments per man, cost per acre, and with the arrangement of labour in various complicated operations. Take for example the threshing of wheat, the carting of hay or corn, the use of the water-drill, or even the carting out of farmyard manure. Labour arrangements form a capital subject for catechetical or oral instruction. Students may be asked to arrange a set of men, boys, and horses for threshing, drilling, grinding, chaff-cutting, pulping, or hay-making, and it must be allowed that knowledge of this description will be of high value in after life, whether communicated in the parish school to boys who at some time will be labourers, or possibly rise to the position of foremen or bailiffs; or in higher grade schools where the recipients of knowledge may be land-occupiers, land-owners, or land-agents. In any case no candidate could for one moment be considered to be an adept in the principles of agriculture if he could not at once describe a process, and allot a proper position to every man or horse occupied in the proceeding.

Costs again offer an excellent field for teaching, as may be exemplified by such instances as an elaborate and well drawn up statement as to the cost of maintaining a farm-horse for one year. Such calculations are most useful. The case cited for example lies at the very foundation of the costs of tillage, and of the cost of producing a bushel or an acre of wheat. We all know the interest which attaches itself to the cost of production, especially in these days of foreign competition, when estimates are constantly placed before us in the newspapers showing how cheaply the Canadian farmer, the Indian ryot, or the Russian peasant can raise a quarter of wheat, and how easily they can displace us in our own markets. Statements will also from time to time appear more or less inexact and misleading as to the cost of producing wheat in this country. Such calculations must be based partly upon materials which can be absolutely valued, as for example the cost of seed, or the cost of manual labour in reaping; but it is not so easy to affix proper value to such acts as that of ploughing, harrowing, or any other operation in which horse labour as well as manual labour is employed. The cost of maintaining a horse lies at the foundation of all estimates, not only as to the cost of growing wheat, but as to the cost of conducting farming operations, and it would be well worthy of the attention of an agricultural teacher that his pupils should be able to prescribe such dietaries for horses at all periods of the year, and to attach prices so as to enable them to come to a conclusion as to the cost of food, and after that to add the cost of shoeing, harness, risk, interest on money, depreciation, &c., and so arrive at the cost of maintenance for a year. Afterwards by an exercise of judgment as to the number of working days in the year, to come at a just conclusion as to the cost to the farmer of his working horses *per diem*. The subject of the rotations of crops has been already touched upon, but it may be well to impress the importance of this section of agricultural knowledge a little deeper upon the attention of my readers. The theory of crop rotation involves not only the question of exhaustion of the soil, but many other considerations. A rotation of crops ought to be so constructed as to conduce to cleanliness or freedom from weeds; it ought also to be so contrived as to give a regular supply of food for live stock at all periods of the year. It ought to conduce to regular employment of horses and of men, and its wonderful power of restoring to land its wholesomeness for sheep ought not to be lost sight of. Every good farmer knows as a matter of experience that sheep will thrive better the first time they are folded over a piece of clover or rape than they will at a later period of the same season when again carried over the same area. The ground becomes stained or sour, and, although the effect may not be injurious upon cattle or older sheep, yet lambs soon give indications that they are not doing so well as could be wished. If it were not for the intervening corn crop we should find our soils unwholesome for sheep stock. If there were no other reason for continuing corn-growing upon our chalk and limestone soils, the restoration of a healthy condition with reference to sheep would in itself be an ample reason for continuing the growth of cereals. The endless variety of rotations, and the varying proportion which they give of roots, corn, and grass offer an excellent theme for exercise and instruction. The modifications of the Norfolk four-course rotation alone, and the curious manner in which certain crops can be substituted for each other so as to lengthen the interval between similar crops, while at the same time the general character or principle of the rotation is not interfered with, might well occupy some hours.

The foregoing subjects do not by any means exhaust the list which might be produced of excellent subjects for the purpose of the teacher of agriculture. When he remembers the varied processes of the dairy, the many considerations with reference to the management of breeding animals, rearing and fattening, as well as of crop cultivation generally, he need not be afraid of running short of a subject, neither need he be thrown exclusively upon such topics as the action of lime upon organic matter, the formation of double silicates in the soil, the change from dormant constituents into active or soluble ingredients in the soil; and while inculcating topics similar to those above enumerated, he will be more truly teaching agriculture and promulgating information which is known to be important to those persons whose lives are being devoted to the pursuit of agriculture. In treating of this subject upon the wider basis which I am endeavouring to lay down, I would recommend that attention should in the first place be devoted to the soil, and that in as broad a manner as is possible. The origin of the soil, the distribution of the soil in accordance with the outcrop of the various geological strata, the conditions of fertility in soils, including climate, subsoil, texture, aspect, as well as mere available plant food, and the possible occurrence of injurious matters which may interfere with what otherwise would have been a fertile soil. The indications of fertility offer a most interesting inquiry for the agricultural teacher which appears to be entirely unexplored and unappreciated by those who profess this subject. These indications appeal to the five senses, and ought to include the consideration of the geological position of the soil in question, contour, altitude, shelter, depth, colour, tenacity, subsoil, as well as the natural produce, as illustrated in the trees, hedgerows, permanent pasturage, weeds, and growing crops. Botany here comes in as the handmaid of agriculture, for certain trees and certain grassy and weedy herbage may be considered to indicate the quality of soil, some plants growing luxuriantly upon bad soil, others requiring good land for their perfect development; certain species of grasses tell a tale of poverty, whereas others may be employed as arguing a high productiveness. This subject will be subsequently enlarged upon. The methods of land improvement form legitimate subject for teaching. The reasons why drainage is beneficial, the action of drains, and the differences in water economy between soils of free character and those which are naturally disposed to retain the water on the surface. The methods pursued and the effects produced by such an important act as land drainage ought to form an important part of agricultural teaching. Similarly the advantages and the dangers of deep cultivation, the advantages of clay burning, of chalking, claying, marling or liming land, and the secrets of successful tillage. The subject of fertilizers ought to be handled practically and not theoretically as it too frequently is. Students should not be taught that silica is an all-important constituent in fertilizers, neither should they be led to think that because a plant abounds in silica, or abounds in lime, or abounds in potash, that silica, lime, or potash are necessarily the best substances for application. I am often told by examinees that wheat takes up a large amount of silica, and that therefore silica must be applied, or that turnips take up a considerable quantity of potash, and that therefore potash must be applied, all of which is erroneous, and springs apparently from faulty instruction. It is propagating the error fallen into by the great Baron Liebig, to whom we owe so much, when he recommended phosphates and other mineral substances in his so-called wheat manure. Wheat, however, appears to be able to get its phosphates, its potash, and its silica in sufficient quantities from the soil, and what it apparently is most in need of is plenty of nitrates, hence if we want to grow a good crop of wheat we must apply nitrates, not silica and not potash. Again if we want to grow a good crop of turnips we must give phosphates and not potash, and not even is it necessary to give nitrogen. The science of manuring, or the principles which should guide us in the application of fertilizers of all sorts, including farmyard manure, ought to occupy the attention of all agricultural classes, and if this suggestion is carried out, the result will be not only a more valuable instruction, but an increase in the respect of practical men for scientific knowledge. Instead of following the lines which I have endeavoured to lay down, we find that agricultural instruction is narrowed to a mere enunciation of certain somewhat crude ideas conceived by theorists with reference to chemical changes which are alleged to take place, but which probably or possibly never take place, and which, if they do take place, have little to do with the practice or the control of the practice of agriculture. Among such ideas I would rank the formation of double silicates in the soil, which are, to say the least, doubtful compounds, which possibly do not exist. Again, the struggles or conflicts which appear according to some of our candidates to be waged in the soil are most remarkable—one constituent of plant food ousting another; nitrate of soda acting as a stimulant or whip; lime feeding voraciously upon organic matter; alumina politely handing in recruits to the hungry roots of plants,—all apparently bustle and activity below ground, and not entirely free from party bias, greediness, voracity, friendship, and nervousness, as though the soil were imbued with vitality. The passage of dormant into active plant food is spoken of in language calculated to give the impression of actual awakening from sleep to a sense of responsibility. The manner in which the change from "dormant" to "active" is worked by examinees is very trying to an examiner. It appears to be the sole idea of both teachers and taught in many cases, and is made to do duty in the case of every question. In the most recent examination of the training colleges the student was asked to describe the manner in which climate influences the productive powers of the soil, and the answers usually contained references to dormant and active constituents. He was asked for indications of fertility, and as an indication of fertility we have "active" constituents. He is asked the principles upon which the effects of irrigation depend, and he says "a great quantity of the dormant matters in the soil can be changed into active." Questions relating to dung, to lime, to tillage operations, to rotation, to climate, to fodder crops, to drainage and all forms of land improvement are answered by repeating this formula as to active and dormant plant food. So impressed is it on the minds of students that they often neglect to further describe them than as active and dormant, as "it turns the dormant into active." This is teaching and learning in a groove with a vengeance. Farmyard dung is treated of in a manner which would astonish most practical men. Its temperature is to be taken; if too high it must be watered, if too low it must be livened up with pitchforks; all of which is outside and out of sympathy with the practice of even the best farmers.

No better guide as to the line which ought to be taken by an agricultural teacher can be mentioned than the contents of the *Journals* of the Royal Agricultural Society, which have been published from the year 1840 down to the present time. An examination of the index of the Transactions of this Society will show the vast range of subjects which properly come under the ægis of agriculture. The varieties of all cultivated plants; new forage plants; the rise, progress, and development of all our domesticated breeds of cattle; the points of excellence in animals; the processes of fattening; experimental results obtained from the use of various kinds of foods; the diseases of plants and of animals, form a very important series of papers. The action of fertilizers, home and foreign agriculture, the reports of the chemist, the botanist, the veterinarian, papers on drainage, on land improvement, papers on statistics, on machinery, on rotations, papers on dairying, on cottages, on buildings, all help to show the vast variety belonging to agriculture as a subject, and put to shame those who reduce what they call the principles of agriculture to a few crude notions with reference to the constituents of plant food, the action of lime, the minute composition of milk, the process of germination of seed, the changes which take place in the ripening of grass or of straw, the action of lime upon the soil, and various other similar topics, which, although they should be included in a system of teaching, ought only to take their place as very secondary when compared with the larger topics which I have endeavoured to put before my readers.

Lastly, I would say that the following chapters are addressed in the first instance to those who have made agriculture their subject as professors or teachers of the same. I have not scrupled to point out generally the errors into which examinees are apt to fall, and I have also endeavoured to keep up the idea of width or breadth as peculiarly the attribute of the great art and science of agriculture. The syllabus of the Science and Art Department is equally comprehensive, and requires a knowledge of all the subjects which have been mentioned in the previous pages.

CHAPTER II.

Agriculture as a Subject—Defective Treatment of it—Proper Function of the Agricultural Teacher—Study of the Soil.

I HAVE endeavoured to show that agriculture may be regarded as a subject capable of being taught, that is, capable of being treated of in the school or lecture-room, always bearing in mind that the pursuit of agriculture is eminently practical, and therefore, although capable of being treated of soundly in the school-room, it would be a mistake to imagine that either as regards its practice or its theory it can be taught by word of mouth alone.

Looking at the subject of agriculture as a whole, I find that it is composed firstly of the processes of agriculture which are capable of description and explanation. Secondly, of varieties in practice, as between north and south, east and west, county and county, which may be properly spoken of as "comparative agriculture," in itself a subject worthy of attention. Thirdly, the history of agricultural progress. Fourthly, the axioms or established facts known to be true by all thoroughly good farmers. Fifthly, the statistics of agriculture, upon which I will not at present enlarge. Sixthly, the economics of agriculture, which take us into the domain of political economy, a subject treating extensively of rent, tenure of land in various parts of the world, and the relations which subsist between the landlord, the tenant, the labourer, both towards each other and to the community at large. I have sketched out in broad lines the topics or sections which make up the subject of agriculture as it ought, I think, to present itself to the mind of an agricultural teacher.

But I have also warned my readers against a mistake which is only too common, and shapes itself to me somewhat as follows—that there is a strong disposition on the part of those who teach agriculture, or who come before us as agricultural professors, to treat this large subject in a narrow manner. I would describe that treatment as molecular and microscopic, rather than as bold and comprehensive. We have agriculture presented to us as a molecular and microscopic subject. When they treat of the soil they enter at once into the constituents of plant food, the action of *bacteria* in the processes of nitrification, the formation of double silicates, the change from what they call "dormant" to "active" constituents. When they mount a step higher, and treat of vegetation, they fix their attention upon *diastasis* and the changes which take place in the germination of seed, or in the growing plants,—upon the action of chlorophyll, or the peculiar functions of stomata in leaves. Or, when they ascend still higher into the domain of animal life, they discourse upon the relative size of the liver or of the lungs in pedigree stock as compared with unpedigree stock, the functions of saliva and other juices in digestion, or the sources of heat, of fat, or of muscle, the albuminoid ratio, or the digestion co-efficient.

However worthy of attention and however interesting these topics are, it must be patent to any one that the agricultural teacher who deals with the minute composition of the soil is really teaching agricultural chemistry, not agriculture. Those who go into the minute changes which take place in germination, or in the absorption of gases by leaves, by stem, and by flower, are invading upon the domain of vegetable physiology, and those who bestow more than a passing attention upon the secretions and functions of the body are constituting themselves teachers of animal physiology. I endeavoured in my first chapter to impress upon you that agriculture is not a patchwork of all the natural sciences, but that it is in itself a great subject upon which the various natural sciences shed their rays of light.

The proper function of the agricultural teacher is to thoroughly explain agriculture upon the broad lines which I have laid down; at the same time to be so alive to the bearings of science, or rather the various branches of science, that he can show to his students the points of contact; that he can point out to them the vast importance of a knowledge of chemistry, botany, and animal physiology; and thereby impress upon them that they are engaged in the study of a truly great subject. If he can do this, if he can make himself the master of agriculture as a subject, and use the natural and abstract sciences as explanatory of his agricultural teaching, he will be more truly useful in communicating knowledge than by entering upon the explanation of the peculiarities of the digestive system, or the changes which take place in the ripening of crops or the elaboration of plant tissues.

I shall endeavour in the following short treatise to at least open this great subject, endeavouring to steer clear between too great attention to any one point, and what might be called unbecoming brevity.

As already mentioned, the subject of agriculture, in the first place, requires a knowledge of the soil. With reference to this part of our subject, we must for a moment consider what soil is. In order to do so, we must imagine the face of the earth entirely divested of animal and of vegetable life, and we would then have revealed to view the bare ground from which all animal and also all vegetable life has sprung. It was an old myth of the Greeks that Aphrodite sprang from the foam of the ocean. However fanciful such an idea may appear, it is absolutely and literally true that all animated nature, all animal and vegetable life, has sprung from the soil. The nature of the soil is therefore a subject well worthy of the pen of my illustrious colleague, Professor Huxley, the author of *Lay Sermons*, and especially of one upon a "lump of chalk," and of another upon "mud." An equally effective lay sermon could be written upon "soil."

Now, the nature and the origin of the soil are subjects of very deep interest. One point I would at the outset lay stress upon, namely, that the soil, as a substance capable of nourishing and supporting vegetation, must have been preparing for thousands, if not millions of years before vegetation was planted upon it. Preparations evidently must have been proceeding for long ages before soils such as we at present see around us could have been produced. The origin of soils is a geological question, but if we ransack works upon geology we shall find but scant attention given to the subject. It is usually dismissed in a paragraph. And, even in books upon agricultural chemistry, the subject is generally all too briefly noticed. We may take it, therefore, that it is left to the agriculturist to study the origin of the soil he tills. I therefore shall discuss at some length the subject of the origin of soil.

The ordinary explanation given as to the origin of soils is the gradual crumbling down and disintegration of compact rocks, under the operation of various existing natural forces, carried on over an incalculable period of time. These forces are, in a large degree, atmospheric, using the expression in a fairly wide sense. Changes of temperature, for instance, which are perpetually taking place between night and day, as well as between summer and winter, produce alternate contraction and expansion, and substances which are exposed to alternate contraction and expansion are apt to break or disintegrate. They are apt to crack up, especially when pressure is brought

ENGLISH AGRICULTURE.

to bear upon them to prevent the requisite molecular changes.

Atmospheric moisture has a powerful solvent effect, no substance, perhaps, having a more uniformly dissolving effect than pure distilled water. When we add to the effect that of carbonic acid gas in solution, then we have a most potent cause of disintegration of rocky material. We cannot have a better reason for the crumbling down and disintegration of compact soil than when its "continuity" is destroyed;. and that continuity is destroyed when a certain portion of the component ingredients is dissolved out. If we take, for example, a granite, and dissolve out the soluble alkaline portion, then the quartz element and the silicate of alumina, which are integral parts of the granite, break down, and are carried away probably by the action of running water. If the soluble part is washed out, then we have the crumbling down of the remainder.

We shall next notice that peculiar change brought about when water assumes the solid state, which it does at the freezing-point. The change is accompanied with an increase of volume which no amount of force is able to check. If porous stones such as sandstones become charged with water, then when frost sets in there is an expansion of the water, and a disruption of the particles. This is known to be a cause of disintegration not only of rocks, but of buildings; and it is the cause of the obliteration of inscriptions on tombstones, and shows itself in very many other ways.

In the next place, we have the wearing action of water, which must be considered in connection with the entire water systems of the world. We have the wearing action of water, not only as exemplified in runnels or cascades upon the mountain-side, but in the wearing action of the entire river system of the world; and of all the currents and tides of the ocean. These forces are in perpetual action, generation after generation; and so potent is the wearing action of water that Sir Charles Lyell, in speaking upon the spherical form of our globe and the reason why it possesses its present shape, does not hesitate to say that, if the globe had originally been turned out angular or cubic, the action of water alone would in process of time have worn down its angles, and the rotary action would itself have so distributed the worn materials that it would at length have taken the form which it now presents.

Neither must we omit glacial action. If we consult Professor Geikie's excellent treatise we shall find, in reference to soil, that he dismisses the subject by attributing the loose matter which covers the surface in a great measure to glacial action. So powerful is this agency that it might well be accredited with the process of soil formation. It is a geological fact, that the whole of Europe and our own country were at some distant period covered with ice, just as we see the Polar regions and Greenland covered with immense thicknesses of ice at the present day. Those who have studied glacial action know well that a glacier moves. A glacier is not stationary, but there is a slow movement where the nature of the ground admits of it—a slow downward movement. Ice is plastic. The particles move one upon another, slowly, but surely. This flowing of the glacier is carried down to the bottom, and the consequence is that the glacier grinds the rocky surface upon which it rests, absolutely scarifying it, absolutely tilling it, if we may so speak, causing great grooves in the direction of the movement. We can still find upon our hills the *striæ* or striated marks of glacial action; and those of us who have had the good fortune to have stood upon a glacier and to have watched it, will have noticed that from its tail, or lowest point, issues perpetually a milky stream, the milkiness or turbidity being simply caused by rocky material which the glacial action has forcibly worn, pounded, and grated; and this material, suspended in the waters of the melting glacier, is carried down to the valleys, and deposited, forming a level and fertile valley—witness the valley of the Rhone.

I have endeavoured to describe the action of the natural forces which have gradually given us the loose material that covers the surface of the planet. It would be a mistake indeed to imagine that soil is merely pounded rock; or that by putting rocky material through a mill we would obtain soil. Two conditions are necessary before pounded rock can be converted into fertile ground. These two conditions are, first, time, and secondly, vegetation. Before we can have a soil such as I speak of, we must not only have certain forces disintegrating and breaking down rock, but ample time must be allowed.

First, then, with reference to time. If we can imagine a sample of soil to be before us, we should observe that it is made up of fragments of various sizes. It contains a large number of mineral fragments, and very little reflection will show that these fragments form a descending series as to size, from pebbles and stones of varying size, down to minute chips requiring the magnifying-glass to render them apparent. Now, these mineral fragments are parts of the original rock from which the soil was derived, and they are all continually exposed to the same cosmic forces which, in the first instance, broke down the continuity of the parent rock. And if we fix our attention upon those minuter fragments which might require the microscope to make them evident, we shall have no difficulty in understanding that there must be a perpetual passage from a condition of compactness to a state of solution. There is a gradual and constant disintegration, and therefore a constant passage from minute but compact material to a soluble state; and this continuous process is what we refer to when we speak of the constituents of the soil changing from a dormant, or, as it may well be termed, a "potential" condition of plant food into what is designated an "active," but which is better spoken of as an "available" condition of plant food. This requires time. Let us then fully appreciate the fact that we have in a soil, as we see it to-day, a certain proportion of soluble matter; that is, of plant food, or mineral constituents in a state in which they can be dissolved by water, and take their place as constituents of the sap of plants.

Next in order, we have an exceedingly pulverulent and impalpable material in all soils, which cannot be produced by the pestle and mortar. We have in the second place—for we have done at present with the soluble materials of soils—a large mass of impalpable smooth material in such a finely-divided condition that it feels "greasy" to the touch—not "gritty," but greasy. This, in a word, is "clay." It has been produced during the lapse of vast periods of time, assisted by the continued operation of forces already named. It is largely composed of silicate of alumina. It gives a peculiar adhesiveness to soil, and it has certain properties which will shortly occupy our attention.

In the third place, we find distributed through the finer portions of the soil an abundant stock of coarser material, from sand to large stones, composing the bulk of the soil whose potential value has already been insisted upon. The general structure of the soil may be briefly summed up as follows—first, a sufficient, but very small amount of soluble plant food; secondly, a quantity of finely-divided, pulverulent material which is useful in holding the soluble matter, and produces a moist condition of the mass suitable for the healthy development of the root, and also all grades of compact material fitted for the recouping of the fertility of the soil. These different states of matter sufficiently indicate that the soil is a product of many complex forces, aided by the lapse of time.

In the next place, before a soil can be fertile, it must contain vegetable matter. Originally, soils must have been without it; but all fertile soils contain the remains of the roots, stems, and leaves of former generations of plants, and, in the case of agricultural land, farmyard dung and other organic substances added to the soil. Without vegetable matter most soils would be unfertile. Vegetable matter gives a consistency to soils most welcome to plants, and confers physical properties upon them which renders them a safe and good anchorage and home for the spreading roots. It regulates the conditions of the soil with reference to both heat and moisture, keeping it cool, for it very quickly radiates heat, but likewise keeping it moist, because it retains moisture with great persistency. One of the chief advantages of repeated dressings of farmyard dung is that the hygroscopic properties of the soil are greatly improved. It enables the soil to retain moisture, it keeps it cool, and it gives it a peculiar texture suitable for the ramification or free passage of the roots of growing plants. These are the mechanical functions of organic matter.

But it has other important uses. Organic matter is the chief source of nitrogen in soils. The vexed subject as to the sources of nitrogen must be looked upon as under controversy. I have no doubt that plants obtain a certain amount of nitrogen from the air. Probably in the long run soil has received it all from the air; but in a cultivated soil, where nitrogen exists in organic matter, no doubt the plants take it largely from the soil. There is, indeed, full proof that nitrogen is taken largely from the soil; and as the organic matter in soils contains large quantities of nitrogen, it is a principal source of this most important element of fertility. By its constant decay, organic matter becomes a source of carbonic acid gas; and as the rain falls upon a soil impregnated with organic matter, it absorbs carbonic acid gas, and its solvent action upon mineral matters is thereby much increased.

CHAPTER III.

Study of the Soil (*continued*)—Its Complicated Character—Soluble Material—Fine Impalpable Material—Quartzoze Material—Calcareous Material—Vegetable Matter—Stones—Complex Functions of Soils—Conditions of Fertility—Indications of Fertility.

IN the last chapter I described the action of certain forces which have slowly caused the crumbling down of rocks, and the difference between a soil and disintegrated or powdered rock. Two factors in the formation of soils account for this difference, the first being that of time, and the second being that of vegetation. It is the lapse of time which produces that extraordinary gradation in the condition of the materials forming soils which may be exemplified (1) by the presence of soluble matter ready to be taken up by the rootlets of plants; (2) by matter in a pulverulent and impalpable condition; (3) by matter in a somewhat coarser condition; and (4) by means of mineral fragments which may be described as portions of the parent rock or rocks from which the soil was in the first instance derived.

This diversity of condition in the material confers upon the soil a special power—a power which is neither strictly physical nor yet strictly chemical, but which has been termed physico-chemical, a power of surface attraction which may be compared to that of charcoal, spongy platinum, and other porous materials, which condense gases and absorb substances from solution. It is a power which has been compared to that of textile materials to fix dyes in such a manner that they cannot be again washed out. Such power is a property of porous substances generally, and is shared by soils. All fertile soils possess it, and by it they are able to retain fertilizing matter for the use of plants.

Further, it is extraordinary that this power is exerted with the most marked effect upon those substances which crops require in the greatest abundance—namely, ammonia, potash, and phosphoric acid. A soil compounded in the manner described is able to seize upon soluble matter as it is produced by disintegration slowly acting upon the mineral wealth of the soil, and to retain such material and hold it diffused throughout the bulk of the soils. Thus the soil becomes charged or permeated with matter which can be taken up by the roots of plants.

Let us now turn our attention to the extraordinary action of vegetation as a factor in the formation of soils. I remember being struck by a passage in Mr. Oliver Wendell Holmes's *Autocrat of the Breakfast Table*, in which he noted the way in which nature is ever ready to place a germ or seed wherever there is a crevice however minute to receive it. She takes possession of rocky surfaces long before they are fit for cultivation or to bear what we call crops, even of grass. Long before decay has actually broken down the rocky texture vegetation places her earliest outposts, and thus helps to expedite that degradation of material which ultimately give us soil. The lichen which tempers and softens the colour of a newly-built wall, the brown or varied colour which we rely upon to tone down the gaudy appearance of a newly-built house, is nothing but the advance guard of vegetation. So also is the moss which requires to be cleaned off the roofs or to be removed from inscriptions; or the ivy which creeps over walls, while beautifying them hastens their decay. Vegetation has many methods of attack, but the result is always a more rapid decay. The mycelium of a fungus or the roots of higher plants penetrate and erode the surface, and the result is the breaking down of the continuity of the substance, facilitating the action of further destructive agencies.

Neither is this all. The presence of these forms of vegetation fosters a condition of moisture favourable to decay. And further, as time goes on, and generations of plants succeed each other, we have the death and decay of plants with the evolution of carbon dioxide, or carbonic acid gas. This gas enters into solution with rain-water, and helps still further to dissolve out the lime and the alkalies, and thus we have in vegetation itself a very potent force in bringing about the gradual formation of that loose material which we call soil.

This concludes what I have to say upon those forces which have during long ages operated in the production of soils.

I propose, in the next place, to deal with what we may call the proximate constituents of soils, the word *proximate* being used in contradistinction to *ultimate*. I am not going to speak of the ultimate chemical or elementary constituents of soils, but of those substances or groups of substances which together make up the entire bulk of every soil.

In the first place, let us turn our attention very briefly to the soluble portion—that is, the portion soluble in water—and clearly understand that this is the true available mineral plant food, the presence of which is of vast importance as a cardinal element of fertility. There are some nine substances which together form what are called the mineral or ashy constituents of plants. They are readily enumerated, but space will not allow me to do more than name them. They are worthy of our closest attention, and should form an important item in any instruction given upon agricultural chemistry in lectures or in text-books. They constitute the available mineral plant food of crops, and are constantly being drawn from the soil, especially by crops which are sold off the farm. They are continually being drawn or washed out of the soil, and the soil is recouped partly by the continual decay of mineral fragments in the soil, and partly by the application of fertilizing substances directly to the soil. The most important of these constituents of plants may be named upon the fingers. First of all there is phosphorous; secondly, there is potash; thirdly, lime; and then follow soda, magnesia, sulphur, iron, silica, and chlorine. Those are the principal constituents of fertile soils; and if we must add to the list we shall have to pass from the domain of what are called mineral constituents, and include nitrogen, hydrogen, and carbon.

If a soil is rich in phosphoric acid, potash, lime, magnesia, sulphuric acid, iron, chlorine, silica, and in a substance which cannot be classed with them, but which is equally important, namely, nitrogen, then we have, to speak broadly, the chief soluble constituents which are required for the building up of plants. Very frequently silica is accorded a very prominent place by examinees. It is difficult to see for why. No one denies the importance of silica, but it is so abundant in all soils that it is scarcely necessary to parade it before us as a very important constituent of a fertile soil. We may take its presence for granted.

Leaving the consideration of the soluble matter, we have in the next place five proximate constituents of all soils, wherever found. We begin with clay, chemically described as the hydrated silicate of alumina. Clay appears to have been produced by the crumbling down of felspathic rocks, such as granite. Granitic rocks, felspathic rocks, and felspar seem somewhat unpromising raw materials from which to derive clays. The forces already mentioned have acted upon them. The alkalies have been washed out and carried away in a state of solution. The sandy or quartzoze element has been detached by running water, and the silicate of alumina has been left as a plastic mass, seen in its purest forms as porcelain clay, white bole, and pipe-clay. Viewed chemically, silicate of alumina cannot be spoken of as a plant food in any sense whatever. The silica is in chemical combination with the alumina, and alumina never enters into the composition of, and is never wanted within, the plant. The silicate of alumina or clay does not take any direct part in the nutrition of plants.

Nevertheless, a clay soil is usually a rich soil. It is difficult to work, and has been under a cloud of late years on account of the low price of corn. Still clay must be considered a guarantee of richness in a soil. For why? It is because in nature substances rarely occur in a state of chemical purity. There are always impurities associated with them, and such is the case with clay. Pure clay is white, but agricultural clay is red, the redness being due to a very important element of plant-food—iron. The redness of clay betrays at once the presence of oxide of iron. In blue clay there is an admixture of vegetable matter, or it may be of lower oxides of iron—protoxides or protosalts, which speedily are peroxidized when they are exposed to the action of the atmosphere.

Clays are frequently described as marls—the term indicating the presence of lime. We shall not find any agricultural clay destitute of lime. One of the great difficulties in making bricks and tiles is the presence of lime in the form of nodules, and it is scattered through the mass in greater or lesser quantities. We may expect to find lime in clays, and with it the acids which combine with lime. For instance, nitrate of lime, sulphate of lime, carbonate of lime, and phosphate of lime. Clays also usually contain magnesia; and they are invariably rich in potash. It is the silicate of potash which gives that peculiar soapy feeling characteristic of wet clay. So, while clay is itself perfectly barren and incapable of supporting vegetation, as it is associated with these substances it becomes an index to the fertility of the soil.

Clay is then associated with various substances which confer wealth or fertility upon soil, but its mechanical or physical properties are also very important. Clay is highly retentive. Not only can it hold a large volume of water, but it retains it with great tenacity. It also condenses gaseous matter upon its surfaces, by which we mean its internal superficies, not its external surface. It exerts the power which is possessed by substances in a very minute state of division. The amount of surface in a cubic inch of clay is vast, for the smaller or finer the state of division in a substance, the greater the amount of its surface.

Clay parts with its heat quickly, and receives heat slowly, so that clay soils are inclined to be wet and cold. The expression "cold clay" is quite usual in speaking of clay land.

Clay is one of the proximate constituents of all fertile soils. Bereft of clay, a soil would be useless for agricultural purposes. Take the clay out, and we should have a blowing sand, and we should deprive the soil of a large number of its principal mineral constituents.

In the next place, we turn our attention to sand. Viewed chemically, sand is as unpromising as clay. We must however clearly define what we mean by sand. First, there is silicious or quartzoze sand, composed of particles of quartz or silicic acid. It is insoluble in water and acids, and in the case of silver sand it is composed almost entirely of insoluble silica.

Sea-shore sand is generally coloured with oxide of iron, but it is also associated with fragments of shell which confer upon it a certain fertilizing value. Viewed purely as silicious or quartzoze sand it is as entirely barren as clay. A mixture of pure sand and pure clay would therefore be hopelessly unfertile, but a mixture of native sand and native clay would not be barren, because the sand, as well as the clay, is intermixed with materials which entitle it to the name of a calcareous sand. There is likewise micaceous sand—a sand in which we find small sparkling plates of mica. While, therefore, quartzoze sand would be entirely sterile for purposes of plant nutrition, yet when it is calcareous or micaceous in character, we have the importation of fresh materials in the same way as we have the importation of materials into clay. Sand, therefore, in its crude or native state may add to the fertility of the soil.

The mechanical effect of sand is the reverse in many respects to that of clay. Silicious sand has very little retaining power for water. Its absorptive powers with regard to water are *nil*. Expose it to a damp atmosphere, and it will not perceptibly increase in weight. But expose dried clay to the evening air, and we shall find that it will increase very perceptibly in weight. The powers or capacity of sand for holding water as a sponge holds water are very small compared with clay. Its power of transmitting water or parting with water is great.

We have here a substance eminently adapted for diluting or neutralizing the characters of a clay soil. Bring the two together, and we produce a substance approaching very closely to the proper conditions of a fertile soil. A loam is simply sand and clay mixed together. I purposely avoid suggesting percentages, and would warn my readers against any elaborate classification of soils—the classification proposed by Schübler about forty years ago, and which appeared in the *Journal* of the Royal Agricultural Society, has never been adopted. The terms are good if not too rigidly employed; but as for distinctly making a classification based on percentages of ingredients, it is not practicable.

A clay soil is of course one in which clay predominates, a sandy soil is one in which sand predominates; and between these two extremes are all the grades of loams.

The gradations which I would suggest are as follows— (1) clays; (2) clay loams, that is, loams which abound in the argillaceous element; (3) loams, which are easy working and friable garden soils; (4) sandy loams, or soils which are still loamy, but in which sand predominates; and (5) sands. Clay stands at one end of the series; then clay loams; loams in the centre; then sandy loams; and sand, forming a series which will enable us to describe very extensive classes of soil.

Lime is a staple constituent of all soils, and occurs as carbonate of lime, or calcium carbonate, which, among the proximate constituents of soils now under notice, is the first which can be spoken of as a plant food. Clay is not a plant food, sand is not a plant food, but lime can be taken up into the tissues of the plant. There is a great difference in this particular between lime and the other two proximate constituents already noticed. Like sand and clay, lime does not occur pure in nature, but is neutralized by acids which take possession of it as a base, and form salts of lime. Carbonic acid is the principal of these acids. All soils contain carbonate of lime, but with the carbonate we find both the phosphate and sulphate of lime. Associated with lime we get these two important constituents—sulphur and phosphorus; and lastly, in the constant processes of nitrification, nitrate of lime is formed.

Lime is of very great use in the soil, both as a plant food and on account of its physical effects. It may be spoken of as intermediate in many respects between sand and clay. This is proved by the experience of farmers who know that lime renders a clay soil lighter, and binds together a sandy soil. This means, that it is intermediate in character between the two. Applied to stiff clay it mellows it; applied to sand it stiffens it.

A large proportion of the soils of this country have originated from the decay of limestone rocks. All our chalk soils are of calcareous character; so are our mountain limestones, magnesian limestones, oolitic limestones, and our marls and marbles of various kinds. They all yield us calcareous soils; and in order to understand intelligibly and practically the classification of soils, the series already given must be modified by the introduction of the term "calcareous." For example, we may speak of a calcareous clay, or a clay in which lime is a feature. Schübler considered that the presence of from five to twenty per cent. of lime in a clay were the limits within which it might be spoken of properly as a marl.

A calcareous loam is a mixture of clay and sand in which lime is a feature. Lastly, we may have a calcareous sand, or sand enlivened and improved by calcareous matter. Lime possesses a function which neither comes under its properties as a plant food, nor yet as a mitigator of the character of soil. It is its power to facilitate the decay of vegetable matter. This is a well-known attribute of lime in accordance with which we are by order of the Privy Council obliged to bury infected carcases of animals in quick-lime. We are aware that quick-lime exercises a very great effect upon organic matter. Lime by virtue of this property tends to exhaust soils of organic matter. It is the base required in the process of nitrification to seize upon newly liberated nitric acid. The nitrification of nitrogenous matter existing in a state of organic combination is one of the most recent and one of the most important facts brought out by agricultural chemistry. Lime plays an important part in this process by promoting the oxidation of nitrogenous matter, and then furnishing a base for combination with nascent nitric acid. Nitrate of lime is easily washed through the soil, and a great deal of it passes away in solution to the lower strata of the soil, or is discharged by drain-water.

Vegetable matter is a proximate constituent of all fertile soils. Looking at it broadly, it is of vast importance, but it is of all the constituents of soils the least absolutely important. It is the least indispensable of the proximate constituents of soils. It is the only one, in fact, which could be actually done without. Vegetable matter could not have existed primarily in the soil, because it is the decayed remnants of the produce of the soil. It is evidently therefore imported. Lava soils are very fertile, although newly formed; it may have been only ejected from the volcano a few years previously in a molten state, and bereft of all organic matter. It is therefore clear that organic matter scarcely can be regarded as so indispensable as the other materials which have been mentioned.

Vegetable matter is sometimes spoken of as an indication rather than a cause of fertility. But it is a cause of fertility as well as an indication. It may be regarded rather as an indication of fertility than as a cause, but it is a cause of fertility notwithstanding. It is an indication of fertility because a soil abounding in vegetable matter has proved itself capable of supporting a vigorous vegetation. A soil which grows a large amount of leaf and of stem must grow a large amount of root. There is also the fall of the leaf and the restoration of the haulm and stubble, which all help to give vegetable matter to the soil. Such store of vegetable matter argues capacity for growth. It shows that the soil is capable. But vegetable matter is a cause of fertility as well as a mere indication, because it contains nitrogen. All organic matter, all remnants of animal or vegetable matter, contain nitrogen. Therefore the organic matter in the soil is a guarantee that it contains nitrogen. It is the cause of the great richness of pasture grounds. Such grounds abound in organic matter, and they abound in nitrogen. One is the index of the other. Organic matter is also a source of carbonic acid gas. As a source of this fertilizing gas in the soil, it becomes with the aid of moisture a digester of mineral matter which otherwise would be much slower in being liberated. Besides, carbonic acid gas is a plant-food in itself. It is taken up by the roots, and elaborated in the tissues of the plant, the carbon being retained and the oxygen being given off by chemical decomposition. Organic matter also contains mineral matter. Straw and haulm and roots of plants cannot be produced without mineral matter, and it is evident that the mineral matter is in precisely the state in which it can be taken up by plants as it is liberated through decay. That organic matter contains mineral matter is a fact which might readily be overlooked.

Organic matter is in a constant state of change—oxidation and decay; and in that decay not only is nitrogen liberated but salts are also liberated, so that there are important chemical changes going forward which help us to understand the great value of organic matter. Experiments have shown that the most recently added vegetable matter is more valuable than the older stocks of similar material which abound in all fertile soils. Nitrification is principally carried on upon the roots, leaves, haulm, farm-yard manure, and other remnants of plant life belonging to the last crop or the most recent dressings. Previous accumulations, owing to their becoming more permanent or fixed in character, are comparatively inert, and although they are sources of nitrogen, they cannot be compared in this respect to substance in a rapid state of decay such as those above mentioned.

With reference to the points that appeal directly to the farmer's instincts, we have in organic matter a wonderful improver of the texture of soils. The organic matter confers upon the soil softness, mellowness, and the power of retaining moisture. It dries very slowly, it gives a coolness to the soil which helps, perhaps, more than any other constituent to make it a wholesome home for the roots of plants. This, perhaps, we will be more thoroughly convinced of if we burn off, as we readily may, the organic matter in a soil. We will then find what a loose, sandy, unpromising material is left. The mechanical action of humus makes a soil a comfortable home for roots to live in and to search for their pabulum, and that is not the least important of its uses. Garden soils are proverbially rich, and they abound in vegetable matter. If I were asked for a good broad indication as to fertility of soil, I should say, "Look at its colour." I prefer a brown, dark-coloured hazel loam. The blackness or hazel colour is due to organic matter. So, we see, organic matter is of great importance, though not of the absolute importance which the older chemists attached to it. They thought that humus was the very key to fertility, but Liebig demolished the humus theory some forty years ago. He showed clearly that humus might be regarded rather as a consequence of fertility than as a cause. Liebig rode his hobby too hard. It was left afterwards to others to point out that, while it is true that organic matter is rather an indication of fertility than a cause, yet it is a cause, and a very potent cause, of fertility in soils.

Lastly, in this particular connection, we come to stones and the coarser and finer mineral fragments. I asked the question in the most recent examination of the Science and Art Department, "Are stones useful in soils, and why?" I was in several cases answered with a categorical "Yes," or "No." Others told me that stones were useless in soils, because "the seed tumbled from one stone to another, so that it got so deep that it could not get up again!"

Now, mineral fragments are useful in the soil. They are, as has been already pointed out, of all sizes, from microscopic chips to big land-fast boulders. Stones are useful because they are constantly being broken down; even tillage implements assist in this. Small pieces are continually detached from them. In addition, there is the constant action of the atmospheric forces completing the passage from the insoluble to the soluble state. Mineral fragments are no doubt a potential source of plant food, and a guarantee against ultimate exhaustion, especially where the stones are parts of the original parent rocks—as, for instance, where there are fragments of chalk in a chalk soil—fragments of the lias clay in a clay soil—fragments of oolitic limestone in an oolitic soil. They are guarantees for the future keeping up of the fertility of the soil. It was the late Dr. Voelcker's opinion that the ultimate exhaustion of soils was a "bugbear." He considered that it was an impossibility, and he was right. Ultimate exhaustion of the soil is an impossibility on account of these fragments. To effect the temporary exhaustion of a soil is very easy. It is done by a quick removal of the soluble portions; but soluble portions are replaced by fresh materials brought gradually out of these stones through attrition, and the very slow action of disintegrating forces, especially upon the minuter fragments. Neither must we forget that stones have a wonderful effect upon the mechanical texture of soils. Of that I could give many examples. Upon the oolitic hills of Gloucestershire, where I resided for many years, we should have had the soil utterly unfit for sheep but for the presence of large numbers of fragments of the underlying great oolite. Bits of the rock break up in flat, plate-like fragments as the ground is ploughed, and the intermixture of these portions of oolitic rock so temper the soil that the farmers are able to treat it as a light soil, and grow barley and fold sheep upon it in the winter.

I have met with many similar cases. Take, for example, the soils round Dartford in North Kent, which is naturally a most difficult clay, so hard that when sunburnt it resembles fine-grained pottery, but which is nevertheless cultivated upon the principles of light-land farming. And why? because it is filled with flint stones—so completely crammed with them that it approaches the character of an open, breaking, friable soil. There is therefore not the least doubt that stones in a soil have a very great effect upon its mechanical texture.

Then there is a third point which depends upon the particular nature of the stones. They may conserve moisture, or they may tend to dryness. Sand stones, or porous stones, which are capable of absorbing moisture, probably give out their store in droughty periods to the surrounding soil; but there may be other stones which have a reverse action, so that we cannot say absolutely which effect they may exert; but we may say that they exert an effect upon the degree of moisture and coolness in the soil. The case is then perfectly clear, that mineral fragments are of use in soils.

We have now worked upwards from soluble and invisible ingredients through the smallest particles to the largest fragments in the soil; and we perceive that we have in the soil a very wonderful and complex substance. I was once struck by a remark made by my much respected friend, the late Mr. John Chalmers Morton, in which he summed up the qualities of the soil in the following words—"It is a storehouse; it is a laboratory; and it is a vehicle." Those are three excellent expressions to bear in mind. It is a storehouse of plant food. It is the grazing ground for plants. It is a physiological and chemical laboratory, in which changes take place of all kinds—changes in the inorganic materials; changes in the organic matter, assisted by countless myriads of bacteria, causing a kind of fermentation which results in what is called nitrification—actions and reactions constantly going on. The conclusion is, that we have in the soil the very womb of all life. It is the nursing mother of us all, and the more we study its wonderful and complicated structure, and the cosmic forces which act upon it and develop its powers, the more will we be struck with the immediate and strong ties which attach all animated nature to our mother earth. This is where all life is elaborated, assisted, or acted upon, of course, by other external forces which must not be forgotten —moisture and a suitable temperature. There must be moisture; there must be suitable temperature; and these, brought to bear upon the wonderful properties of the soil, cause the germination of the seed and the growth of the plant. These changes are closely and inseparably connected with the growth of animals; they are equally and inseparably connected with the existence of man; with the functions of brain and of thought, with poetry and art, and all the higher developments of mind, all of which actually spring out of the soil. What is wanted before these results can be realized is the creative fiat, or the planting of the seed. The principle of life must be assumed in the first instance, but once place the fertile germ in the warm and moist matrix, and expose it to the action of air and sun, and the rest follows in a manner analogous to the development of the embryo in the womb.

The great primitive forces of fire, air, earth, and water, which Aristotle called the elements, are all engaged, and all that is needed beyond them is the fertilized germ or seed, the presence of which we accept, but cannot account for except by the simple statement that God planted it.

Two subjects require to be carefully separated and defined by pupils studying under the Department. One of these is the conditions of fertility in a soil; the next is the indications of fertility.

By the "conditions of fertility" we mean the parallel to what is meant by "the conditions of health." The conditions of fertility in soils are parallel to the conditions of health in animals. The conditions of health are very different from the indications of health; and the conditions of fertility are very different from the indications of fertility. To carry the analogy a little farther, I should say that the conditions of health in a man or in any animal might be enumerated as follows—plenty of good food, plenty of pure air, plenty of occupation, plenty of rest, a healthy constitution inherited and kept up, &c. These are some of the conditions of health. But the indications of health would be a well-nourished body, a healthy complexion, vigour as indicated in action, and due powers of endurance. But we should not for a moment dream of confusing the conditions of health with the indications of health. And yet after some years' experience of this particular point which I have tried to illustrate, I find that, after the student has answered the question on the conditions of fertility, he repeats himself when asked for the indications of fertility. He answers, "Plenty of plant food." That is not an indication at all; it is a condition.

Let us address ourselves to the conditions of fertility in a soil, and ascertain what are the conditions without which no soil can be fertile, and with which a soil is always fertile.

The first condition of fertility is, as you might expect, plenty of plant food in an available state. A student may make two points or one of it. A great many students like to make as many points as they can. It looks well, they think, to see fifteen points made instead of five. Plenty of plant food, and of course in an assimilable form, is a condition of fertility. But there is a great tendency to regard this as everything—that is to say, to take an exclusively chemical view of the matter. Students are too apt to think that if a soil is well stocked with plant food in an available state it is a fertile soil, but such is not the case. The wonder is, how little mere quantity of plant food has to do with fertility.

At one time it was thought to be all-important, and I do not doubt its importance; but there may be a very small amount of plant food, and yet a fertile soil if the plant food is in such a state that the plant can readily get at it. On the other hand, we may have a rich soil in which the tenacity of the soil or other circumstances make it very difficult for the plant to get at the plant food. The plant food may be locked up. Hence chemical analysis does not always give us such a ready key to the fertility of a soil as we might expect.

A great deal depends, as already pointed out, upon the ease with which the roots can pursue and capture their food, if I may so speak, all of which is not very easily indicated in a sample of soil. In such a case we cannot tell very much what the exact conditions of the soil are. We read the analysis, but require to know a great deal more before we can come to a conclusion as to whether it is a fertile soil at all. Nevertheless, we shall put as the first condition plenty of plant food in an available form.

What other conditions of fertility are there? A very important one indeed is that the soil should be deep, because a thin soil of three or four inches only in depth, though a rich one, may not be half or a quarter as good as a soil in which there is a depth of a foot or a couple of feet. There must be quantity, in fact, as well as quality, and you have a guarantee of quantity in the depth. In addition to a high percentage of plant food, I say, we must have depth of soil. This is, then, a second condition of fertility.

A third condition of fertility is the absence of directly injurious substances from the soil, such as sulphate of iron or green vitriol, and sulphide of iron or iron pyrites. These may be directly injurious substances in a soil. Among these are also substances of a sour character, as, for instance, partially decomposed vegetable matter. It is a condition of fertility that such substances should be absent.

Fourthly, the soil must admit of the ingress of air, the free passage of water, and the free passage of roots. The texture of the soil must be suitable in all these respects.

The soil must rest upon a suitable substratum, and if it does not we cannot have a fertile soil. We must have a suitable subsoil. That is of vast importance, and ought to be included in all instruction upon the true conditions of fertility.

Then there is climate. I am aware that climate is an external circumstance, but it is most important. I mean climate as influenced not only by latitude, but also by longitude, altitude, aspect, and shelter. Climate must be taken into account before we have all the necessary conditions.

Now I believe that we have treated the subject before us fairly exhaustively—that is to say, if we have a soil in which there is an abundance of plant food in an available state, in which there is plenty of it as indicated in depth of soil, not only a percentage, but an actually big amount of it, if it is free from injurious substances of all sorts, if we can have it of proper texture, if it is lying upon a suitable subsoil, and if it is exposed to a suitable climate, we have the six conditions of fertility. A soil with all these conditions is a fertile soil; a soil without any one of these is not a fertile soil. Some of these conditions are under our control, and others are not. Those within our control may be produced by possible improvement. Others of these conditions, such as climate, for example, are generally, but not altogether, out of our control.

As to the indications of fertility, I shall only open the subject. I must introduce them, if only to contrast them with the conditions of fertility. When I ask for the indications of fertility, I very generally elicit what I have just been enforcing as the conditions of fertility. The indications of fertility are those features which appeal to the unassisted senses, and which point out to the skilled observer the character of the soil. The indications of fertility are so evident that the man who has his attention drawn to them may judge land to a certain extent out of the railway carriage window at express speed. For instance, the general contour of the country gives a fair presumption about land. The indications of fertility and of barrenness are very numerous, and no indication can be taken as conclusive in itself. They must be taken each in connection with others. The contour of the country is an indication of fertility. Precipitous broken ground is not promising; sometimes fertile nooks and corners occur, but a mountainous, rocky, precipitous, highly picturesque country is not likely to impress the agriculturist with a sense of its fertility. But, on the other hand, gentle slopes, and especially flat tracts, give presumably an idea of a fertile country; especially is this the case where we have what are called alluvial tracts, either extending from the coast or accompanying the course of rivers.

I have often come to the conclusion in rapid travelling that land is stiff or light by the mode in which the ridges are laid out. If we see round-backed ridges about three or four yards across we shall be impressed with the idea that we are going through a stiff land district; but if, on the other hand, the land is laid out in wide ridges, we are probably going through a light land district. I mention this because I have spoken of certain indications which are to be noticed in a very cursory inspection.

A well-wooded country is generally a fertile country. We must not expect to find well-grown timber on chalky downs or open wolds. A tree is there a scarce article, and it is very often a weather-beaten, poorly developed specimen. At the homesteads and on the village green it is different. Homesteads are generally fertile, even in districts of poor soils; and the village green is rich through its long association with human habitations. On fertile land we shall find elms, oak, ash, sycamore, walnut, mulberry, and hawthorn, all growing luxuriantly, attaining a great size; and we cannot then have much doubt that the ground is good. On the other hand, a puny class of timber gives an opposite impression. Do not, however, be led away by large beeches, because the beech is capable of growing upon a poor thin soil; such beeches as may be seen at Blenheim, or in Gloucestershire at Oakley Park, are splendid samples of timber, but still it is beech. Beech will attain its height and size, but is not any indication of fertility. There are also many conifers, such as larch or Scotch fir and spruce, which grow to a large size even upon poor soil. They are, in fact, a great boon to the owners of poor soils. There are also birch, varieties of poplar, blackthorn (in contradistinction to whitethorn), and alder. All of these trees will grow upon very poor and wet soils, and although the list might be extended, it is sufficient for my purpose to draw attention to the fact that good timber is an indication of good land, while puny timber is the reverse. At the same time, we must take into consideration the sort of timber, and by an examination of this feature we shall be assisted in our judgment. Hawthorn has been already mentioned. There is an old saying among practical men, that " bad land never grew a good thorn." Good thorn hedges are a very excellent indication of the fertility of land.

We will take, in the next place, as a further natural feature, the permanent pastures. Permanent pasture is an excellent indication of fertility or the reverse, because it is often unsophisticated. In most districts permanent pasture remains much as it was in the beginning, whereas arable land has either been improved by good farming or has suffered by bad farming. Permanent pasture is a good indication of natural fertility. Good permanent pasture is of a rich and lovely green. That peculiarity is of the greatest value as an indication of quality in early spring. All pastures are green in summer unless they have been burnt up by drought, but in early spring we like to see the pastures green. The reverse may be expressed as follows—they are bleached, white, or bluish green instead of a vivid, beautiful green, or they are red and rusty looking. An abundance of sedges, daisies, and "moons" (that is, ox-eyed daisies) is a very bad sign. So is an abundance of "Dothering Dick" (*Briza media*), Yorkshire fog (*Holcus mollis*), and barren brome (*Bromus sterilis*). All of these are indications of either poor or wet land. Pasture of this description, separated by a hedge from a good crop of wheat, shows the true character of the soil.

A spongy, wet, soaky feeling experienced in walking over land is also a bad indication. Dig a turf and see whether the mould is deep, or whether it is thin-skinned. Judges of land do not like a thin-skinned pasture, but deep, black soil and plenty of it. Thus as an indication of fertility or barrenness an inspection of the pastures is of great importance.

In the next place, let us take the crops. I have purposely deferred their consideration. I am sometimes told "that good crops are indications of good land, and bad crops are indications of bad land; and even that rich farmers are indications of good land, and poor farmers are indications of bad land; and that good, well-built farm-houses, and good, well-built buildings indicate good land, and badly-built houses and badly-built farm-buildings indicate bad land!" That is a style of answer I sometimes receive. Well, all this is true in a sense, but it is nevertheless rather wide of the mark.

With reference to crops, they are not a very good indication of the fertility of land. As a general truth, good land will bear good crops, but good land sometimes bears bad crops when it is badly farmed; and bad land very often bears good crops, but it is done at great expense. Good crops can be grown on both bad and good land, but at a greater outlay. Perhaps two root crops are taken in succession, and then a fine crop of wheat is grown; but on good land it would grow equally well without the two root crops. Now this is a matter less of natural fertility than of what is called *condition*. The expression is well understood by land agents and by good farmers when it is said that land is in "good condition" or in "bad condition"; that is, it has suffered or benefited from management to which it has been subjected for several years past. The success of a crop depends upon the condition of the land as much as it does upon natural fertility. While, therefore, we are insensibly pleased with good crops, we must be on our guard about them. To answer a question as to indications of fertility by saying that good crops indicate good land and bad crops indicate bad land is both inexact and misleading.

Weeds are a very good indication. I have been told that freedom from weeds is a sign of good land, but it is not so, not in the least. How can cleanliness or freedom from weeds be an indication of good land? It is a mere indication of care on the part of the cultivator. Several sorts of weeds are an indication of goodness in land—chick-weed, stinking mayweed, nettles, really big strong milky thistles, the great Scotch thistle growing to a big size and to a great height, are all good indications. I remember in the valley of the Theiss, in Southern Hungary, being struck with the size of the thistles. I saw thistles of such portentous size that the men mowed round them instead of cutting them down. We do not like to see couch, but if it is there, we prefer its being fine and large, strong and vigorous, not wiry, puny, and benty. Poor benty couch, or agrostis alba, canina, or nigra, or others of the agrostis tribes, are bad signs on land. There are certain weeds which are indications of good land or the reverse. There is a certain number of weeds which are mentioned as signs of good land, and others which are called signs of bad land; but unfortunately we can generally find the weeds that ought to grow on bad land on good land, and the reverse. We are constantly finding weeds that are described as denizens of the one class of land upon the other; but the real point is, that it is not so much the description of weed as the manner in which it grows. Great, early, and vigorous growth is a sign of quality in land, and a puny, badly developed growth of natural herbage of any sort is a sign of poor land.

Bracken (*Pteris communis*) is a good sign. Heath is a very bad sign. So we see that the natural herbage may be profitably studied by the land valuer with a view to giving him a sound judgment.

CHAPTER IV.

Indications of Fertility (*continued*)—The Subsoil—Geological Position of Soils affecting Fertility—Geological Knowledge to be used with Caution—Geologically Recent Soils.

I HAVE endeavoured to explain the difference between the conditions of fertility in soils and those indications of fertility which actually are the basis of land valuation, or, to speak more correctly, the grounds upon which the land-valuer arrives at his conclusions as to the quality, if not the value, of land. And in comparing the conditions of fertility in soil and the indications of fertility, I mentioned that the latter were evident to the senses; that is to say, that the contour of the landscape, the character of the trees and of vegetable growth of all descriptions, as well as the colour, depth, and other peculiarities of the soil, all came under the head of indications of fertility. But there are some indications of fertility which are not recognizable by the senses, as, for example, the composition of the soil as arrived at in the laboratory. The analysis of the soil I look upon as an indication—one out of many—of the comparative fertility of the soil. Another indication which is only apparent to the instructed mind is that which is afforded by geological position. An analysis forms, when viewed intelligently, one, and the geological horizon or the geological position forms another indication, which ought not to be lost sight of. With these two exceptions, the remainder chiefly appeal to the five senses.

E

In further reviewing the indications of fertility or the reverse, we are able to divide them into those indications which spring from what the soil is capable of producing, being, in fact, an illustration of the old proverb, that the proof of the pudding is in the eating of it. Good crops, good pastures, good trees, all that the ground produces, may be considered as a fair criterion as to what it is capable of producing. And after dismissing those matters, we enter upon the absolute examination of the soil itself.

In doing this the colour of the soil is an important consideration. To speak briefly, rich brown-coloured soils are fertile, although not always so. But in newly-ploughed up land a patchy, a piebald, or varied appearance is decidedly unfavourable. If from a slight eminence the newly-ploughed field has an appearance which may be compared with that of a badly-made cheese—patches of blue, patches of white, yellow, and red—it is a bad indication. Black oxides of iron, silver sand, peat, ochreous deposits—all are indicated by a piebald, parti-coloured soil. And more than that, a parti-coloured soil is never a deep soil, the variety of colour being in a great measure the result of a thin skin or thin character of surface soil. This change of colour will be especially noticeable at the open furrows, where the ground is parted and exposes the subsoil to view.

The texture of the ground is ascertained by the feeling underfoot. Strong land clings to the boots. To carry a pound of clay on either boot in walking over ground is a very different sensation from that caused by walking over light sandy soil, where you walk without soiling your shoes. The texture of land may also be ascertained by handling it, by forming it into pellets, and noticing by the sense of touch and by the sense of sight how far those pellets of soil are plastic and tenacious, or crumbly and sandy. "A rope of sand" we know is proverbially descriptive of a want of tenacity of purpose; but a cylinder of clay taken from a clay soil is very tough, and indicates the tenacious character of the soil.

These are the leading appearances so far as surface soil is concerned.

Next we come to an examination of the sub-soil. It is most important that the soil should rest upon a wholesome sub-soil. By the term "sub-soil" I prefer that you should understand the ground beneath the plough-sole. We might speak of "a soil six feet deep"; but I should prefer saying that the soil and the sub-soil are alike fertile in character. For our purpose as teachers I am sure that it is well to define the sub-soil as I have done, as the section of the ground which lies immediately under the plough-sole.

Of course a sub-soil of equal character—equal appearance and colour and texture with the surface soil—is a most happy combination. It indicates great fertility. It is to be seen in what are called "alluvial" soils. We get there a deep bed of alluvium, in which it is very difficult to discriminate between soil and sub-soil; and in such cases, owing to the sorting action of water, we find both soil and sub-soil rest, deeper down, upon a gravel bed. In all river deposits the stones sink to the bottom and the finer materials accumulate on the top. Hence alluvial soils usually lie upon a gravel bed, and the river which divides them secures their thorough drainage. Alluvial soils are invariably rich. Where soil and sub-soil, then, are of similar character, we have a happy combination. Many good soils rest too immediately upon gravel. Such soils are excellent as long as the seasons are fairly moist, but readily burn up and become brown as a turnpike under the influence of a prolonged drought. Where a good soil rests upon a gravelly sub-soil we have a burning soil, and also what may be called a hungry soil; that is, a soil which too rapidly allows fertilizing matter to pass through it into the deeper and more inaccessible layers of the soil.

In the third place, we occasionally find a clay soil or a stiff loam resting upon an exceedingly tenacious blue clay. Such soils are "holding" in their nature. They will hold fertilizing matter and moisture with great tenacity, and they may be converted into useful fertile soils by the aid of artificial drainage. But in their natural condition such soils are apt to become waterlogged, and consequently unprofitable for arable purposes, although they form a very excellent basis for permanent pasturage.

Sometimes a clay soil or a stiff loam rests upon sand or gravel, giving highly favourable conditions. And, in the fifth place, we find the reverse, namely, a light sand resting upon a clay. That is a favourable condition, not only because the clay assists the sand to hold moisture, but because the marl or clay from below may be dug up and spread upon the surface, a system extensively carried out in the county of Norfolk, and which is also applied to the fen-lands of Lincolnshire, where the clay has been the means of reclaiming or rendering useful thousands of acres of land.

Sixthly, we find soils in which the rock is so near the surface that the plough-sole may be heard rumbling on the top of the rock. Sometimes the plough is arrested by the rock itself, and the horses are stopped. Now, when the surface soil is thin and rests upon the rock, we have a good example of what was stated in a previous chapter, that is, a soil which may give a good analysis, but is deficient in quantity. Of course a great deal depends upon the nature of the rock. A rock of a moist character, conserving moisture, and of a fissured nature, is not unfavourable. It is stated with reference to hops, for example, which are a deeply-rooted plant, that they prefer a soil of this description with a fissured and rocky sub-soil, or rubbly sub-soil, into which the roots may penetrate to a great depth.

In other cases the rock is of an arid character, in which

case we shall find an unfavourable combination. And in some cases, when farmyard manure is applied and ploughed in, it is ploughed down literally upon the rock. If farmyard manure is ploughed down on the rock it probably quickly decays, as air rapidly gains access to it, and the nitrates and other valuable constituents are lost through drainage through the fissures of the rock. This has led the farmers, where such conditions of soil exist, to avoid ploughing in farmyard manure. I have known excellent farmers sow their wheat and wait till it appears above ground, and after it is up cart on the dung at a suitable time, and spread it over the young wheat, thereby protecting it to some extent from the rigours of the weather; and the decay of the dung takes place on the surface, and the fertilizing elements filter gradually down to the roots of the plant. This system is adopted in order to avoid the dung being wasted through the fissured rock.

In the seventh place, a chalk sub-soil ought to be mentioned, as the chalk formation covers a vast area of ground in this country. The upper chalk forms a very excellent sub-soil. Chalk is a source of plant food, and it allows a free passage of the roots and of water down into it. Great benefit accrues from the digging up of chalk and spreading it on the surface. I am sorry to notice that it is done much less frequently now than it appears to have been done in our grandfathers' time. In chalk districts there are evident traces that chalk has been largely used. We can scarcely ride over fifty acres of land without seeing old chalk pits forming depressions in the surface. Farmers seem to have become discouraged by the low prices, and I think that this has been partly the case. They are in no mood for trying expensive and laborious improvements. Change in practice in this particular may likewise be due to the cheapness of special substances, such as superphosphates.

But whatever the cause, there is much less chalking done than there used to be in the old days. A chalk sub-soil is also cool and moist, so that in a long and protracted drought it is pleasant to notice how the thin white soils of the upper chalk retain their verdure.

We must therefore take the sub-soil into account as not only a condition of fertility, but likewise, as it is evident to the eye, an indication of fertility.

Another indication of fertility or barrenness is geological position. Agricultural teachers ought to have some insight into the subject of geology. It might be considered feasible in speaking of the distribution of soils in this country to describe them county by county, and to take the geographical rather than the geological view of the distribution of soils. If we take up the Journals of the Royal Agricultural Society, we will find that the agriculture of every county in England has been described. The reading is exceedingly instructive, whether we take up our own particular county in order to better understand it, or take up the subject for the purpose of comparison. The agriculture of Wiltshire, the agriculture of Oxfordshire, or the agriculture of Northumberland or Lancashire, form very valuable reading. But in all cases the authors, in describing the different soils of their counties, are obliged to resort to a geological classification.

Again, we must remember that knowledge is much more easily retained, and much more easily communicated if it is connected. When facts come to be strung together like beads upon a necklace, that is to say, when there is one thread running through a large number of isolated facts, it is much easier to carry those facts away with us. It is with that object chiefly as connecting various agricultural facts that I introduce geology. It so happens that England can be readily divided and described, by observing the succession of the various geological strata, as the soils of this country succeed each other in a remarkably orderly manner.

Without going minutely into the matter, let me remind my reader that the crust of the earth is composed of a large series of rocks, superimposed upon each other in orderly succession.

We know, for example, that London is situate upon the London clay. The London clay forms "the London basin." Chalk hills rise to the south, and chalk hills rise to the north of it. The Kentish Downs or the North Downs rise south of London, and the hills of Saint Albans or Hertfordshire rise on its north. The chalk, again, is divided into upper and lower chalk. Under the chalk is the upper and the lower greensand. Notice that these likewise form a part of what is called the London basin. Next in inferior position is the Weald clay, which, however, is restricted to a well-known district. The Weald clay occupies an important position in the counties of Sussex and of Kent. It is often spoken of as a clay soil, but it is not invariably so, there being a great deal of light land in the Weald. After that, in descending series, there are the upper oolites ushered in by the Purbeck beds, the Portland beds, and finally by the Kimmeridge clay. Next the middle oolite, and the lower oolite, then the lias, the new red sandstone, the magnesian limestone, the coal measures, the mountain limestone, and the old red sandstone at a vast depth beneath the surface. Such is briefly the geological staircase. If a well could be sunk in Trafalgar Square the boring would in turn penetrate these formations. They are superimposed upon each other in the order in which they were originally deposited, and if we examine a geological section of England we shall find that the above is the order of succession of these rocks. But the point which I wish to call attention to is, that each of these rocks in turn occupies the surface of the country, and it is not altogether easy for a student of agriculture who has not attended

56   THE PRINCIPLES OF

GEOLOGICAL SECTION FROM HERTFORD TO BRIDGEWATER.
(L. B. LOADER'S MAP)

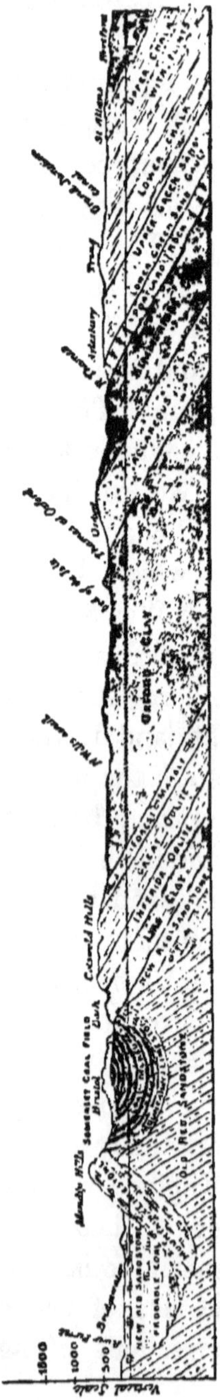

Fig. 1.

geological lectures to see how the new red sandstone, buried as it is under hundreds and thousands of feet of depth, can ever form the surface, as do also the coal measures and the still deeper strata. At the risk of being considered trite I must follow up the explanation. It has been effected by an upheaving volcanic force by which the inclination of these beds has been altered, so that they have been upturned in a manner which can be easily shown by a diagram (Fig. 1). Their edges have been exposed in the way indicated. More than that, we find that large tracts of country, by the process of denudation, have come to be represented by certain rocks which at one time were buried under vast accumulations of deposited matter. Examination of a geological map shows that the upheaving force to which I have alluded has been applied upon the north and upon the west sides of these islands. If we examine the general physical features of Great Britain, we see evidence of this in the north and in the west of England; in Cumberland, in North Wales, in South Wales, in Devonshire, and in Cornwall. There is evidence that volcanic force has raised those parts of the country, giving them a rugged contour and increasing their altitude, the country, in fact, rising in those directions into mountains with other evidences of upheaval; whereas upon the east side of England the country is flat. The upheaving force has therefore been applied upon the north and upon the west, and, as might be expected, the outcrop of the various formations takes place in a north-westerly direction. Now what is the real state of the case? It is this. If we take the train from London to Manchester we traverse first the London clay; we then cut through the chalk; we then traverse the greensand; we next find ourselves upon the soils of the upper oolite, then upon the soils of the middle oolite, then upon the soils of the lower oolite; we shall then cross the lias and the new red sandstone. We shall not cross the magnesian limestone, because it happens to be omitted in that particular route, but cross the coal measures, after which we shall arrive at Manchester. Now these are remarkable facts, and it is of great assistance in coming to a conclusion as to the distribution of soils in this country.

It would take us a long time to go through the peculiarities of the various soils resting upon different geological strata. But let me recommend it as a study—and as a most important one. It has never since been prosecuted with such thoroughness as it was by the late Professor Johnstone of Durham. In Johnstone's *Agricultural Chemistry*, now brought out in a later edition by Dr. Cameron, but especially in the old edition of Johnstone's *Agricultural Chemistry*, is to be found a most painstaking account and comparison between the nature of the soil and the geological formation from which that soil is derived.

There is no doubt a strong resemblance between soils, whether found in one county or another, which have the same geological horizon. If the idea is once grasped that there is a close relationship between the succession of rocks in the geological section of this country and the soils which successively occupy the surface of those rocks, that is the idea which I wish to impress.

Certain formations give soils of a high average fertility, and other formations give soils of a low average fertility. Wherever we have a mixed soil, formed by the blending of two or more formations, there we are likely to have a fertile soil. In other words, the soils which occur at the margins where formations meet together—the soils which occur at the confluence of two formations—are usually of better character than soils which are in the centre, or well into the area occupied by such formations. For instance, soils which are derived from the London clay mixed with the chalk are more fertile than either chalk or London clay soils.

Soils derived from the mixing of lower chalk with upper greensand are usually of high average fertility. Soils derived from the London clay formation mixed with soils derived from the Suffolk or Norfolk Crag are generally richer than soils derived solely from the London clay. The mixing of the rocky debris of two or more formations in the production of a soil is favourable, and very generally gives a strip or region of fertile ground. It is the same principle no doubt which gives the value to alluvial soils. They have been brought down by the river as it cuts its way through a number of different formations, so that the washings of many kinds of soil are brought together, and a high standard of fertility is the result. This is a point of some importance in connection with the bearings of geology upon the distribution of soil.

Let me, in the next place, enforce the importance of care in using geological knowledge as a means of predicating the character of soils. I am aware that the London clay is a tenacious, fine-grained soil. It may be seen to be such on the north of London, towards Harrow, and it is certainly a striking fact that so much ground there lies in permanent pasture. It is remarkable in such close proximity to a vast population, but it is accounted for by the retentive character of the ground, which is best suited for pasturage and producing hay for the London market.

In Essex the London clay reigns paramount, and perhaps no county has suffered more severely from the recent fall in the value of cereals than Essex.

The student is in some danger of rushing to the conclusion that when the London clay forms the predominant element we must necessarily find a clayey soil. The London clay formation is, however, liable to interruptions, and is not exclusively a clay formation. The Bracklesham beds and the Bagshot sands form part of the great formation which is known as the London clay; and where these beds occur we find the very reverse of what might be expected. An intimate knowledge of surface as well as of block geology is needed before a conclusion as to the relation between the nature of the surface soil and the characteristics of particular formations can be arrived at.

Similarly we might make a serious error if we concluded that all the soils upon the lias clay are stiff. The occurrence of sand beds in the lias clay often modifies the soil, and explains the presence of light land even in valleys of the lias. Neither must the student forget the overspreading of a great part of the North and Midlands of England with drifted materials, sometimes many feet in thickness, which have overspread the main geological formations, and given the surface a fresh character. Take, for instance, the new red sandstone, which, as a rule, yields a red soil of high average fertility. That character of the new red sandstone is well maintained in Warwickshire, Worcestershire, Derbyshire, Cheshire, Cumberland, and in Yorkshire. But when we cross the Tees into the county of Durham, although upon the geological map the peculiar colouring of the new red sandstone predominates, the soil alters in character to a poor boulder clay of considerable depth. That boulder clay, or drift clay, masks the character of the new red sandstone. No longer does the new red sandstone actually form the surface, but a cold, intractable, and infertile soil, which spreads over a great portion of the south-east of Durham, and not only masks and covers the new red sandstone, but likewise to a great extent the magnesian limestone adjoining. Accidental features may also be introduced by the occurrence of peat, which may be found upon any geological formation, as, for example, where the Oxford clay is largely covered with peat in East Anglia, and again where, south of Doncaster, and in the neighbourhood of Selby, the new red sandstone is masked or covered by alluvial deposit and mixed marshy and peaty soils.

Again, the course of every river is marked by the deposition of alluvial soil. This again forms interruptions, and alters the character of the soil from what it might be conceived to be by the general student of geology.

In some cases lava floods have been thrown over the surface, and have altered its character. Thus, in South Italy, in the neighbourhood of Naples, the lava thrown from the cone of Vesuvius has overspread the original limestone of the country, and by its decay given a soil of a much richer character than could have been produced by the original dolomite which forms the main feature of the peninsula. In taking up the study of agricultural geology we must be careful, and if the subject is to be studied it must be studied minutely.

From what has been stated earlier in this chapter, we know that the newer formations are to be found upon the east and south, while the outcrop of the older formations is towards the west and north. And in accordance with this rule the largest development of recent soils is upon the east coast*; not that such is exclusively the case, but the largest development of recent soils—of soils of recent geological origin—occurs in the south and east portions of the country. First, and further north, there is that great extension of recent soils which constitute the south-eastern extremity of Yorkshire, known as Holderness, a flat, rich, alluvial tract, which is steadily increasing in area, and which terminates at Spurn Head, the extreme south-eastern point of Yorkshire.

Spurn Head is constantly growing, so that the lights which

---

* In reading the following pages upon the geological distribution of soils reference should be made to the geological map which forms the frontispiece of the book.

are required to guide navigation have to be altered from time to time. Crossing the Humber towards Grimsby in Lincolnshire, we find a continuation of the same class of ground. Holderness is composed of a rich, flat, clay soil, first-rate bean and wheat land, and flat and rich pasturage. Holderness is the original home of the shorthorns, which were first imported from this flat, fertile range. In close proximity to Holderness, in a northerly and westerly direction, the chalk formation rises out of the alluvial plain.

The chalk formation surrounds Beverley, Driffield, and Malton in the East Riding. This is not a populous district. Standing at Beverley, and looking in a south-easterly direction, we see the fertile flat tract of Holderness towards the sea, and behind us rise the wolds of the chalk formation.

Crossing the Humber, we have on the eastern boundary of the county some eight miles in width of the Lincolnshire marsh. This tract forms a fine expanse of fertile marine clay, very similar, indeed, in character to Holderness, and similarly backed on the west by the chalk formation, forming the agricultural district of the Lincolnshire Wolds.

The marshes extend southward to the town of Burgh, not far from the Wash. At Burgh there is a great extension of alluvial deposits, extending inland almost to the city of Lincoln, forming an extensive area around King's Lynn on the Wash, and a considerable portion south of Mid-Lincolnshire, Huntingdon, Cambridgeshire, Norfolk, and Essex. Here are situated the city of Peterborough, the town of Spalding, the town of Boston, and around these centres lies the well-known fen country. Any one who travels north by the Great Northern railway passes over this district in running down to Newark *via* Peterborough. The alluvial soils under notice commence in Yorkshire, forming South-east Yorks, cross the Humber, forming East Lincolnshire, run down to Burgh, then extend inland, and give the large district around the Wash, extending over the counties which have been named.

Now what is the character of the fen country? It is strikingly flat, and in many cases wonderfully fertile, divided agriculturally into marsh and fen, two words which have quite a different signification to a fen farmer to what they might have to an outsider. The marsh is one thing; the fen is another. The marsh consists of a fine laminated marine clay of very high fertility; the fen is composed of black peat capable of producing much straw, but is not so productive in grain.

The marsh is nearest the sea; the fen is more inland. The marsh yields some of the richest land in England, and perhaps the most extensive view of it is obtained from the top of Boston "Stump." From the top of Boston stump, that is, the tower of Boston church, is to be seen a very extraordinary view of perfectly flat country, including in a twenty mile radius some of the best land in England. The fens are less productive. They are black, peaty soils, unpromising, and were at no very distant period the home of snipe and wild fowl, but are now reclaimed. On these fens extensive claying has been adopted, the Oxford clay having been dug up and spread over the surface, so that the character of the soil has been completely altered.

The fen country is interesting, because it has been wrested from the ocean by enterprise. The sea walls which have been erected from time to time are of vast extent and of very massive character, so much so that in driving over the country the road lies along the top of these sea walls. In studying the successive stages by which reclamation has been effected, we find, first, the Roman wall, and upon the land side or inside of this wall are ruins of churches and abbeys. Later in our tour, if we drive along a wall erected in King Charles II.'s time, we may see no ruins of churches, but many old mansions, and a fairly old country with good timber. If, further, we ride or drive along the wall which has been erected in the Victorian period, we shall see a great deal of land, very recently in cultivation, not carrying any large timber, and much of it scarcely yet free of its brackish character. It has been suggested to still further increase this area by taking in the Wash, but enterprise in agriculture has been sadly checked of late years, and we have lately heard nothing of the project. There is no doubt that a great deal of land might be reclaimed from the sea. It is now only ooze, and is covered and laid bare alternately by the tide twice every day. I have endeavoured to sketch the general character of the fen country of East Anglia in order to show that by following the geological order, a certain similarity in soils passing from county to county may be traced. Instead of considering Holderness and the Lincolnshire marsh as isolated and separate from each other, we look upon them as connected, as both belonging to the same geological conditions, as both being in a great measure formed by the action of the river Humber, and also by certain ocean currents which are still acting, and which will in due time increase the amount of fertile soil in those regions.

Having spoken of this particular class of soil, I must in the next place point out the peculiarities of some other geological formations, beginning with the London clay and proceeding with the chalk.

CHAPTER V.

Soils of the London Clay—Soils of the Chalk Formation.

IN the last chapter we traced the area of greatest development of recent soils. We have next before us that group of formations known as the London clay, and I must draw your attention to them as briefly as possible. The London clay formation may be roughly described as occupying an area of triangular shape, the base extending from or near Marlborough in Wiltshire to the city of Canterbury, the perpendicular rising from the city of Canterbury to Saxmundham in Suffolk, the hypothenuse starting from Saxmundham in Suffolk, and extending to or near Marlborough in Wiltshire. Thus we have a triangular area, including the estuary of the Thames. Within that area we find a portion of South-east Suffolk, nearly the whole of the county of Essex, a portion of Hertfordshire, the whole of Middlesex, a portion of North Surrey and Kent, a part of Berkshire, Hampshire, and a small portion of Wiltshire. Such is the area which is occupied by the soils of the London clay, the great central city being London. With respect to the agricultural characters of the London clay, we may say that where the London clay really predominates, there the soils are of a stiff character, as, for example, in Middlesex, and over the greater portion of the county of Essex, and a portion of Berkshire, especially that portion on the south side of Reading. The London clay, when it actually does form the soil, gives land of retentive character, but, as I have had occasion to point out previously, the general stiffness of the London clay is interrupted by the Bagshot series, the Thanet sands, and the Bracklesham beds, which predominate largely in the neighbourhood of Windsor and Epsom, and pass onwards towards Aldershot, giving in that part of the field, soils of a light and poor character.

The valley of the Thames also, with its large accumulation of alluvial soils, in the neighbourhood of London, modifies and alters the general character of the soil, so that in many cases within this area a light soil occurs, and in other cases an unusually fertile soil. Where the London clay comes into close contact with the upper chalk, which it does around its entire margin, a favourable change in the general character of the soil is noticeable. The drainage is improved by reason of the underlying chalk sub-soil, and the admixture of a large number of flint stones with the clay tends to open it up and make it more fit for ordinary agricultural operations.

Another large area covered by the soils of the London clay may be described as follows:—It also is triangular in shape, and upon reference to a geological map, we shall find the apex of the triangle at Salisbury, and the boundary stretching in a south-easterly direction from Salisbury to Worthing, and in a south-westerly direction from Salisbury to Weymouth; the base of that triangle being the sea-coast passing the Isle of Wight, and extending from Worthing to Weymouth.

Although the London clay proper yields strong lands, yet by far the greatest area is composed of light-topped, sandy, gravelly soils, frequently of very poor character indeed, including the whole of the large district known as the New Forest. This interesting tract is of such poor character that it is difficult to believe the truth of those old stories which tell us that the Conqueror devastated a fertile country, and destroyed a vast number of villages, hamlets, and churches. It would scarcely support such a population in the nineteenth century, and what it was in the eleventh century I leave the reader to imagine. The ground is exceedingly poor, sandy, gravelly, and miserable, varied by good strong land around Southampton, where the clay forms the surface.

In the next place, we must take a brief survey of the chalk formation. The first development of the chalk occurs north and west of Holderness, forming the Yorkshire wolds. The wolds terminate at the coast in the bold escarpment known as Flamborough Head, and north of Flamborough Head there is no chalk. Extending inland from Flamborough Head, and bounding Holderness on the north and west, is the well-defined district of the Yorkshire wolds. On its southern boundary the Humber cuts it; but if we take the ferry-boat and cross the Humber, we shall find ourselves upon a corresponding and similar district, namely, that of the Lincolnshire wolds, resting upon the same geological horizon as the Yorkshire wolds.

Standing upon the eastern spurs of the Lincolnshire wolds, we look over the flat tracts of the Lincolnshire marsh, just as looking over the spurs of the Yorkshire wolds we looked over the flat tracts of Holderness. The Lincolnshire wolds run down as far as Burgh, and there they disappear underneath the accumulated drifts of the Wash, and the alluvial deposits which occupied our attention so recently. Now let us pause for one moment in order to describe the peculiarities of the Yorkshire and Lincolnshire wolds. Both are elevated tracts, composed in some cases of precipitous ground, which cannot well be brought under cultivation, in other cases of rounded hills cultivated to the summits. They are well adapted for sheep farming, and also for the cultivation of barley, turnips, and clover crops; it is a fair-cropping, easily-worked soil, divided into large farms and large fields, supporting a prosperous and intelligent tenantry. Such is a general description of the wolds of both Yorkshire and of Lincolnshire—a soil, light, free, thin, but grateful and safe cropping. Where it mingles with the soils of the upper greensand the fertility of these soils is much enhanced. Having disappeared under the accumulated drift it reappears in the county of Norfolk, and the chalk then becomes a continuous formation, which may be described as follows.

It extends in a south-westerly direction through Norfolk, Suffolk, Essex, Cambridgeshire, Hertfordshire, Buckinghamshire, Oxfordshire, Berkshire, and Wiltshire. In Wiltshire it reaches what is called the central plateau of the chalk—that is, Salisbury Plain, and from Salisbury Plain, which we may take as a centre, it sends out the arm which we have just traversed terminating in Norfolk. It sends down another arm through Wiltshire and Dorset, past Shaftesbury and Blandford to the Dorset coast. It sends out a third arm from the north-east into Hampshire, extending to and forming the Kentish North Downs, passing through Surrey and North Kent to Canterbury; and it sends forth another arm to Beachy Head and Brighton, giving what are called the South Downs. Such is the distribution of the chalk, looking at it from the central plain of Salisbury. It sends out one great arm into the county of Norfolk, which we have already traced. But while it sends an arm in a north-easterly direction, it sends another arm in a south-westerly direction, through the county of Dorset, giving the Dorset Downs, and ending at Weymouth; it sends out a third arm from the north-east through Surrey and Kent, right away to the city of Canterbury; and it sends out a fourth arm in a south-easterly direction to Beachy Head, the termination of the South Downs. It lies on the geological map like a great octopus or star-fish, stretching its arms in four directions. Such is briefly the distribution of the chalk. Notice that it is continuous, and that a pedestrian could walk over every part of it without leaving it. With reference to the general characters of the chalk, they are best considered in connection with the different counties through which the formation passes. First, with reference to the chalk soils of Norfolk, we cannot say much, Norfolk soils being the most difficult to classify or arrange of any soils in England. They include all kinds of variations, from a stiff clay to blowing sands and calcareous chalky soils. It would be impossible, in fact, to describe the soils of Norfolk. It is evident that even geological knowledge does not enable us to say very much with reference to the characters of these soils.

In Suffolk it is somewhat different. But there is a considerable similarity between the soils of North Suffolk, and those of South Norfolk. They vary very much, and the chalk, in fact, scarcely asserts its character in these two counties. But when we leave Norfolk and Suffolk and come into Essex, Cambridgeshire, Hertfordshire, Buckinghamshire, and the other counties already named, we find undulating or rolling districts called Downs. Thus we have the Downs of Cambridgeshire, the Downs of Buckinghamshire, Essex, Hertfordshire, Oxfordshire, Wilts, Hants, Dorset, &c. They form the North Downs, which extend past Croydon in Surrey, and the Kent Downs, and the South Downs, which stretch through Sussex towards Lewes and Brighton. Downs are similar in character to what were described previously as wolds in Yorkshire and Lincolnshire. In the north the chalk hills are called wolds, but in the south they are called downs. They are sub-mountainous, undulating, sometimes even approaching the grandeur of mountain scenery, in which latter case cultivation is not attempted. The upper portions of the downs produce a short, sweet, and somewhat scanty herbage, well suited for the peculiar breeds of sheep which graze them, and it is to be regretted that a too sanguine enterprise some years ago broke up large portions of these downs. They can hardly pay for cultivation at the present day, and in the case of many thousands of acres broken up thirty or forty years ago, their owners would gladly see them back again under the greensward; but this is by no means easy. The chalk does not take kindly to grass, and it is a difficult task to get it back again into the state of natural herbage in which it originally was. It is too apt, after growing grass for the first two or three years very satisfactorily, to become "benty," thin, and poor, and to require to be broken up again, and put through a course of husbandry. It will grow grass in alternate husbandry, but it is difficult to restore the character of herbage which originally existed. If space allowed, I could cite cases in which this difficulty has been overcome. I have seen instances in which as good herbage has been produced as the original down. In such cases the problem has been solved; but in others the difficulties have proved insuperable. In describing the soils of the chalk, it is necessary to divide them into the upper and the lower chalk. The upper chalk, which forms the high-lying downs, abounds in layers of flint stones. I have already spoken of the upper chalk as giving a light soil—grateful, easy-working, dry, wholesome, well adapted for sheep and safe cropping—all of which characteristics were mentioned in connection with the Yorkshire and Lincolnshire wolds. The lower chalk gives a distinctly better soil. It is to be found stretching away from the more westerly and northerly spurs of the downs, and extends northwards and westwards from the main line of chalk, and, as might be expected, being older, it crops up northwards and westwards, and forms vales and valleys or flat lands north and west of the chalk hills. It is destitute of flints, much grayer in colour than the upper chalk, and yields soils of much greater fertility.

CHAPTER VI.

Geological Section from Hertford to Bridgewater—Soils of the Greensand Formations—The Gault Clay—The Weald—The Upper Oolite—The Middle Oolite—The Lower Oolite—The Lias—The New Red Sandstone—The Permian Formation—The Coal Measures—The Mountain Limestone—The Millstone Grit and Yoredale Rocks—The Old Red Sandstone.

ALLOW me to ask attention to the geological section from Hertford to Bridgewater (p. 56). I have introduced this section to bear out what was stated in a previous chapter, viz. that the inclination or dip of the strata which underlies this country is from north-west to south-east, and that consequently the outcrop of the strata is in a north-westerly direction, that is to say, we shall find as we traverse the country from the south-east to the north-west that we in turn pass over the strata in the order of their outcrop. The diagram speaks for itself, and it will be unnecessary to devote more space to its explanation.

With further reference to the chalk formation, notice that almost the whole of it is inhabited by particular descriptions of sheep, namely, the Down breeds. It is true that the chalk wolds of Yorkshire are chiefly under Leicesters, and the wolds of Lincolnshire are principally under Leicesters and improved Lincolns; but as soon as we cross the Wash and come to the main chalk commencing with the county of Norfolk, we find a hardy, black-faced, short-woolled ovine race known as Norfolk Downs. There is also upon the Suffolk hills a Suffolk Down sheep of excellent quality, and upon the Essex Downs there is a recognized Essex Down. If we follow out the chalk counties successively we shall find that each county has its Down breed of sheep, some being more conspicuously known than others. I may mention Oxford Down sheep (a crossed race), Wiltshire Down sheep, sheep of the North Downs or Kentish Down sheep, sheep of the South Downs or Sussex Downs, sheep of the Hampshire Downs, and sheep belonging to the Dorset chalk district, known as the Dorset horned sheep. All of these sheep, with the exception of the last-named, may be described as brown or black-faced and shanked, with somewhat close or short wool, thriving well upon a dry upland, and preferring a short and wiry dry pasture to a profusion of long and coarse grass, such as is found in lower grounds or valleys. Another peculiarity of the soils of the upper chalk is that the hills are very frequently capped with a better class of land than the slopes and subsidiary valleys. This is well known to land valuers, and appears to be caused by the admixture of the soils of the upper chalk with the soils of the London clay, the most of which have been removed by denudation, but which still to some extent cap the rounded summits of the hills of the upper chalk. The tops of the hills of the chalk are accordingly often of stronger and better character than their flats and slopes.

Having sketched out the general pathway of the chalk, the next formation which must occupy our attention is the greensand. The greensand may be spoken of as forming a ribbon or closely applied band upon the north and west boundary of the chalk formation, bearing out the rule which we have found to hold good in the distribution of our soils. This sinuous ribbon of green, shown on the geological map, which is known as the greensand, will be found to be properly divided into the upper greensand and the lower greensand, the two being divided from each other by the gault clay. Glancing at this formation, we find that the upper greensand is as a rule a fertile soil, while the lower greensand is barren.

The two offer a great contrast to one another—the upper greensand giving fertile soil, and the lower greensand giving not only barren, but almost completely worthless and sterile tracts.

The lowest portions of the lower chalk rest upon the upper portions of the upper greensand. The junction usually occurs upon the north and west flanks or slopes of the chalk hills. There we may expect a singularly excellent soil, whether in Cambridgeshire or in Bedfordshire, or at almost any point within the line; at Devizes, Warminster, or Shaftesbury, for example, the soils are of exceptional fertility. The gault clay is a formation which does not cover a large area. Where it does appear it gives a soil of great tenacity, exceedingly expensive to work, and not very promising for agricultural purposes, especially at the present day. The gault clay, however, only forms a comparatively small portion of the surface of this country, its greatest development being in the county of Kent in the neighbourhood of Sevenoaks, where a great deal of the characteristic gluey clay of this formation occurs. The lower greensand frequently extends as moorlands carrying heather, stunted and coarse herbage, and underwood. It is composed largely of silver sand, and of other sandy loose material, which, although useful to gardeners for potting purposes, has little true agricultural value. There is a large extension of the soils of the lower greensand in North Kent around Godalming, Dorking, and south of Maidstone, as may be seen in travelling from London to Hastings, or from London to Brighton, when no one can fail to be struck with the barren character of the districts through which he passes.

The next geological formation is the weald of Sussex and Kent. This forms a thoroughly well-defined agricultural district, and it may be described as partly consisting of heavy clay, and partly of soil of a lighter character, according as the actual weald clay predominates, or the soils which are derived from the Hastings sands and other sandy formations associated with the Weald, such as the sands of Folkestone, Sandgate, and Hythe. No doubt a large portion of the district is exceedingly difficult land, more suitable for pasturage than for arable cultivation, and of a nature which requires a very considerable amount of skill in order to render the cultivation profitable. It has earned rather an evil notoriety, a good deal of money having been sunk there in unprofitable farming.

I will next ask attention to the large and extensive series of soils, which are based upon what is known as the oolitic formations. These soils occur upon the north and west boundary of the greensands. Accordingly they are to be found in the first place in the county of Yorkshire, forming a considerable portion of the East and North Ridings. They extend in a south-westerly direction, occupying a large portion of the Midlands of England, and trend gradually towards the mouth of the Severn, so that from the mouth of the Tees upon the north-east to the estuary of the Severn on the south-west we find a broad area of the country covered entirely with soils belonging to the oolitic formation. It is necessary to divide this great section in the following manner —in the first place the upper oolite, in the second place the middle oolite, and thirdly, the lower oolite; and in looking at these three series of the oolites we shall find that the upper division is composed largely of the Purbeck beds, Portland beds, and Kimmeridge clay. The two first-named formations yield soils of light character, while the Kimmeridge clay forms soils of clayey character, but of fair and even good quality. As the Purbeck and Portland beds are restricted chiefly to Dorsetshire, we find that by far the larger portion of the upper oolite is composed of Kimmeridge clay. The Kimmeridge clay is in fact the predominating element in the upper oolite, a point which ought to be remembered, and which will be brought out as important presently.

Next, turning to the middle oolite, we shall find that it is ushered in with certain thin limestone rocks, which are known as the upper calcareous grit, followed by the coralline oolite, followed again by the lower calcareous grit, and again by the Oxford clay and Kelloway rock. Those are the main rocks which compose the middle oolite; but, like the last formation, the calcareous grits and coralline oolite are rather exceptional than general, and the Oxford clay occupies a large proportion of the area occupied by the middle oolite. So much is this the case with reference to both the upper and the middle oolite, that frequently the Kimmeridge and the Oxford clays alone represent these two formations, passing together in a southerly and south-westerly direction, and giving a tract of clay soil upon the north and west of the greensand formation. And as the greensand is itself only narrow, this compact bed of clay seems to pass with little interruption from the north-west flanks of the chalk formation. This is highly interestingly shown in the county of Lincolnshire, a county in which the geology of the district corresponds in a striking manner with its agriculture. Let me briefly recall the distribution of the soils in Lincolnshire in this connection. Upon the extreme east is a district of marsh, then westward of the marsh are seen the chalk wolds passing southwards to Burgh. Westward of the chalk is a district which is known as the clays, the district which is now occupying our attention; the clays lie upon the west side of the chalk, so that in Lincolnshire we have successively from east to west marsh grounds, chalk hills, clays, and finally, although this is anticipating, the Lincoln heath; those four—marsh, wold, clays, and heath—passing from north to south of the county, and dividing it into four well-defined strips.

Resuming the consideration of these two formations, we may first notice them in Yorkshire, forming the valley of Pickering, and the district of clayey land west of Scarborough. In the next place we shall notice it in Lincolnshire, giving the clay district already mentioned. Passing a little further south, past Peterborough, it forms a district of strong land in Huntingdonshire, around Kimbolton and Huntingdon, and passing right under the fens, giving the basis for the peaty growth of the fen country. This is the clay which is dug up and spread over the surface of the fens. It gives the clay district round St. Neot's, Olney, and Bedford, the clay districts of Buckinghamshire, the clay district of Oxfordshire, from which it takes its name; then it passes into Wiltshire, yielding a well-defined clay district round Swindon and Cricklade, all of which forms a flat vale easily seen from the neighbouring chalk hills looking from them in a north and westerly direction. Next it passes into Somersetshire, and there becomes broken or masked by alluvial deposit from the river Severn. We have now traced the upper and middle oolite chiefly as illustrated by the Kimmeridge and Oxford clays. The Kimmeridge clay is agriculturally superior to the Oxford clay, which is of poor character. The Oxford clay gives a very poor soil; the pastures even being poor. In the county of Gloucester it is almost uncultivated. In that county it forms Braydon Forest, a large and extensive district, but hopeless for the purposes of cultivation, although it forms a happy hunting-ground for sportsmen.

We have in the next place to very briefly examine the soils of the lower oolite, and it is necessary to divide them approximately, at least, as they are very various in character. The lower oolite is composed in the first place of the rocks known as corn brash, which in turn rests on the forest marble. Then we have the great oolite, the Stonefield slate, the Fuller's earth clay, and the inferior oolite

With reference to the soils of the lower oolite, the corn brash is restricted in area, but yields a fertile soil. Wherever found it yields an excellent soil. It is always associated with bands of clay, the result of the decay of the Bradford clay and the forest marble. Next comes the great oolite, which gives a light thin soil well adapted for sheep-farming, but of moderate or inferior fertility. Underneath it is the Fuller's earth clay, which again yields a soil of retentive character. Below this clay lies the inferior oolite, the lowest member of the group giving free soils of cold inferior character, rising up into rather precipitous heights, and where cultivated giving sheep farms. Taking the lower oolite as a whole, from the corn brash to the inferior oolite inclusive, it usually constitutes a high-lying district, or series of districts, rising up north and west of the clays of the upper and middle series. Thus in Yorkshire it forms that picturesque district known as the Cleveland Hills, a mountainous district lying to the north and west of the Vale of Pickering, extending from Whitby and overlooking the vale of Cleveland. The Cleveland Hills are in the majority of cases uncultivated, but they have extraordinary mineral wealth, chiefly in the shape of jet and ironstone. The Cleveland Hills form an important feature in the North Riding of Yorkshire; they are traceable past Thirsk. The formation reappears on the southern banks of the Humber, and traversing West Lincolnshire in a southerly direction, forms a wold-like hilly tract known as the Lincoln heath already mentioned. We have no longer the bold outline of the Cleveland Hills, but a cultivated tract, on which there is a fairly good light soil well adapted for sheep-farming, and known as the Lincoln heath. The lower oolite forms a considerable portion of the county of Northampton, but in that county it has not the distinctive character which it exhibits either in Lincolnshire or Yorkshire. After it passes through Northamptonshire, where it is the basis of excellent grazing country, it reasserts its bolder and more mountainous outline in the county of Oxford, gradually rising into the district known as the Cotswold Hills. It forms the Cotswold Hills in Oxfordshire, in Gloucestershire, and to a small extent in Wiltshire and Somerset. In the hills of Oxfordshire and Gloucestershire we have a repetition of the bold mountainous character seen in the North Riding of Yorkshire. The Cotswolds are in some cases uncultivated, being too precipitous for arable cultivation, but in more numerous cases yielding good sheep land, either in original down, or under the plough. The same character of soil as is found in Oxfordshire is also to be found in Somersetshire, where the soils of this formation predominate.

Following up our survey of soils, I have in the next place to ask attention to another extensive tract of land lying on the lias clays. In geological order they come below the inferior oolite, and they are divided into the upper, middle, and lower lias. The lias clay is applied closely upon the north and west margin or boundary of the lower oolite, forming a succession of valleys. The lower oolite forms hills and bold escarpments with deep indentations, the effect being highly picturesque. The first illustration of this is to be found in the Cleveland Hills, where we look from the spurs of these hills in a northerly and westerly direction over the fertile valley of the lias known as the Vale of Cleveland. The same rich character of land is found traversing South Yorkshire until it passes into Lincolnshire, and then it bends through the Midlands, giving land of high average fertility, vale-like in character, and frequently in permanent pasture. Such is the character of the ground round Melton Mowbray and Market Harborough, forming the well-known hunting country of Leicestershire, which owes its high character for hunting chiefly to the fact that it lies so much under permanent pasturage, affording every opportunity for sport. The strength of the land is such that it requires very superior horses and excellent horsemanship. Still more important to us as agriculturists is the fact that we have there a fine dairy district; the rich deep pastures of Leicestershire are well known in connection with the manufacture of Stilton cheese. Next we pass into Warwickshire, Oxfordshire, and Worcestershire, and further south into Gloucestershire, where the lias clay soils form three extensive vales well known to all agriculturists of the south and west as the Vale of Evesham in Worcestershire, the Vale of Gloucester, in which the city of Gloucester and the town of Cheltenham are situated, and the Vale of Berkeley—three extensive valleys of rich fertile land, extending from the north flanks of the Cotswold Hills. There is, in fact, abundant resemblance in landscape and general features between these valleys and the Vale of Cleveland in Yorkshire, all of them extending in a northerly and westerly direction from the outlying spurs and escarpments of the lower oolite. The valley of Stroud is another example of the same thing, the valley being deeply cut and narrow, resting upon the lias clay, and surrounded by the hills of the lower oolite.

It has been remarked by Mr. Bailey Denton, who is a great drainage authority, that there is no clay north of the lias. That of course we cannot receive as absolute, but when we travel north of the lias no more extensive clays are to be met with, and so far we may agree with him that the last of the great clays of this country have been noticed, these clays being first the London clay, secondly the Weald clay, thirdly the Kimmeridge clay, fourthly the Oxford clay, fifthly the lias clay; and beyond them there is no other great or continuous clay band which can be traced geographically and geologically.

The next formation is the new red sandstone, which again may be described as applied immediately and at once to the northerly and westerly boundary of the lias clay. I have sometimes in describing the distribution of soils all over the country compared it to one of those dissected maps which is sometimes used as a toy for children. A dissected map of England might easily be made in which the workings of the fret-saw follow the sinuous outline of these formations, and these might be pieced together much in the same manner as a child's map of the country for the purposes of instruction, so completely is the junction between the successive formations followed out. With such a fret-saw map before us we could place upon the northern and western boundary of the lias clay our pieces in such a manner as to convey a correct idea of the succession of the new red sandstone. The new red sandstone generally rises into hills, and as the lias forms valleys at the foot of the lower oolite, so the observer generally looks across the valleys of the lias to the hills of new red sandstone on the north and west.

As the lias forms the valley which is overlooked from the spurs of the lower oolite, and as the new red sandstone forms hills on the far side, the view reveals a vale or valley of lias resting between the oolite hills on the south and east, and the new red sandstone upon the north and west. The new red sandstone is generally characterized by soils of fertile character. When the new red sandstone formation has fair play, that is, when it really is the parent of the surface soil, we have a red-tinted soil of high average fertility. The first illustration of the new red sandstone occurs in North Yorkshire and South Durham, upon the north and west boundaries of the lias clay.

This is the portion of south-east Durham which has been already mentioned as being covered over by a deposit of drifted material which has completely disguised the character of the new red sandstone, and given the soil an inferior character on the north bank of the Tees; but in Yorkshire it gives highly fertile soils south of the Tees. It then passes through Yorkshire—York itself occupying an important central position, and in mid-Yorkshire the soils are spoken of as being of a mixed character, varying from clayey soils to blowing sands.

In South Yorkshire the new red sandstone is masked by alluvial deposit, which alters its general character. In the county of Nottingham it forms an excellent agricultural tract, which contrasts favourably with the neighbouring coal-fields, and it is here that four ducal properties meet, known as the "Dukeries." The city of Nottingham is the most southerly point of the Yorkshire and Nottingham coal-fields, and at Nottingham the new red sandstone formation suddenly increases in width, forming a considerable portion of the Midlands of England.

This formation gives soils of a high average fertility in the county of Derbyshire, contrasting favourably with the locality known as the Peak, and forming the district known as the Garden of Derbyshire; it yields a fine agricultural tract in the county of Staffordshire, superior agriculturally to the potteries and coal-fields of the black country.

In Leicestershire it furnishes a fine tract of agricultural land of better quality than the soils of the coal-fields of the county. In Warwickshire it produces that remarkably fine fertile country round Warwick, Leamington, and Kenilworth, known as the "Heart of England." I suppose there is no finer district for the tourist and agriculturist than this. There we have an exceedingly fertile soil, partly due to the new red sandstone, and partly due to the lias clay. In Worcestershire, also, we have a fine tract of ground, lying upon new red sandstone, adapted for orchards, and once more we find the hop-plant thriving in the county of Worcestershire and up to the city of Worcester. In Shropshire the new red sandstone yields soils of excellent character, and in the county of Cheshire it gives us the foundation for the well-known cheese-making and dairying districts of that county. In Lancashire the new red sandstone also gives some tracts of excellent land.

We see then that wherever the new red sandstone formation extends we have land of good or promising character, and it is worthy of notice that its soils are of much higher quality than those of the numerous coal-fields which lie scattered at intervals over the Midland counties.

There is an extension of the lower red sandstone of the Permian further north—in Cumberland, where it forms a very excellent district round Carlisle and Penrith, extending northwards to the shores of the Solway Firth.

The formation which is next in order is that of the magnesian limestone, which terminates at the city of Nottingham. It follows in close juxtaposition the Yorkshire coal-field. The magnesian limestone occupies an area which may be described as extending from the city of Nottingham right through Yorkshire, and finally developing into a wider area in the county of Durham. It runs from the city of Nottingham almost due north through Yorkshire, and its greatest development is to be found in the county of Durham.

The whole of the coast-line between Hartlepool and Shields is composed of magnesian limestone, and extending inland it forms a considerable portion of the county Palatine. There is no other portion of the country that lies upon magnesian limestone. These rocks usually rise into low hills, carrying a thin soil of poor character. Magnesian limestone or dolomite is yellow or cream-coloured, and is very different in appearance and properties to the mountain limestone, a formation which we shall have to consider presently. In the county of Durham the same drift which masks the new red sandstone covers up and masks the magnesian limestone, so that the soils of the magnesian limestone of South Durham are generally replaced by boulder clay, and soils of somewhat inferior agricultural value.

In the next place, we must notice a large class of soils in association with the coal measures. The coal-fields of this country come in geological order below the magnesian limestone; they are therefore associated with the magnesian limestone and with the new red sandstone. The principal coal-fields are the Northumberland and Durham, the Yorkshire, the Lancashire, the Staffordshire, the Forest of Dean, the Bristol, the Leicestershire, or the coal-field of Ashby de la Zouch, the Warwickshire, the great South Wales coal-field, and others.

These coal-fields are a source of great national wealth, and have an important agricultural value, because they give us markets for our produce. The black spots on the geological map will be found to contain the names of most of our great manufacturing centres. Within the limits of the Lancashire coal-field we have Wigan, Chorley, Warrington, Blackburn, Oldham, Bury, Preston, Rochdale, Manchester, Birkenhead, and Liverpool; in the Yorkshire field we find Bradford, Leeds, Sheffield, Halifax, Wakefield, Burnley, Huddersfield, Rotherham, Staleybridge, Chesterfield, and Nottingham; and again we may instance Birmingham, Wolverhampton, Bristol, and Cardiff.

These coal-fields—or black countries—give us an unrivalled market for our produce, but it is a singular fact, that for agricultural purposes they are not of the highest value; they are not particularly favourable for agricultural pursuits, and that for more than one reason. In the first place, in the coal-fields agriculture ceases to be a first-class occupation; it is overshadowed by the greater mineral wealth which lies beneath the surface. In the second place, there is such a large population, such a large number of roads, and such an immense amount of traffic as to render it rather difficult to carry on agricultural pursuits with the same ease and pleasure with which they can be carried on in more rural districts. In the third place, the air becomes polluted. It is melancholy to read Mr. Clare Sewell Read's report on farming in South Wales. He speaks of blighted trees and blighted crops, owing to the gases vomited forth from a thousand chimneys; and cases are not uncommon of damages having been claimed and recovered in these districts for injury caused to vegetation. All of these things are against agriculture; but, in addition to this disadvantage, the soil is usually poor. Nature, which has been so lavish in underground wealth, seems to have held her hand with reference to the surface fertility of these districts. They are often composed of cold soils, and taking into consideration all their features, they are not favourable for agricultural purposes.

I have yet to mention the mountain limestone in association with the coal-fields. Our great national wealth is in great measure due to the proximity of limestone to coal.

The mountain limestone almost invariably surrounds the coal-fields. Whether we take the Bristol, the Northumberland, the Yorkshire, the Durham, or the Welsh coal-fields, we find that the mountain limestone is never far distant.

The mountain limestone gives soils of poor thin character. They, together with the rocks of the millstone grit and Yoredale rocks, unite the Yorkshire and the Lancashire coal-fields, giving a somewhat barren succession of moorlands intersected by fertile dales; and, in fact, from the Peak district of Derbyshire, where the mountain limestone commences, through the dales of West Yorkshire and East Lancashire, right up to the moorlands which surround Stanhope, Kilhope, and Walhope, and the Weardale district of South Durham, and down into Northumberland, where we find ourselves upon the moorlands, extending to Allen Heads and Cumberland, pastoral pursuits prevail and agriculture

ENGLISH AGRICULTURE. 85

has ceased to exist. The domain of agriculture has ended. I remember when I was a youth studying farming in mid-Northumberland looking over these moors and heaths, and being told that I might walk without meeting a fence from that portion of mid-Northumberland down to the Peak of Derbyshire. The tourist could pass right down through the moorlands of Durham into the dales, moors, and uplands of West Yorkshire, and might actually continue all the way down until he came to Dovedale and the Peak district of Derbyshire; so that when we take into consideration the coal and manufacturing districts of Yorkshire, and the manufacturing and coal districts of Lancashire, and the fine extent of dales and pastoral lands from the Peak of Derbyshire up to Northumberland, it is clear that so far as a survey of agricultural lands is concerned we have come to the end of our task.

There is one more formation which has not been named, and which must not be omitted, and that is the old red sandstone, which comes up below the mountain limestone from the depths beneath. The old red sandstone, as might be expected, is to be found upon the far west and north; for instance, it first comes to the surface on the north bank of the river Severn, forming a portion of Monmouthshire, Herefordshire, and a small part of Worcestershire. It crosses to Brecknockshire, and very generally gives soils of extraordinarily high fertility. Like many other formations, it is divided by geologists into upper, middle, and lower. We may dismiss the upper and lower as giving districts in which the soils are remarkably poor, as, for instance, in Sutherlandshire and in parts of Gloucestershire. But the middle members, or the cornstones and the marls, give soils of high fertility, so that the old red sandstone is spoken of as highly fertile. This is so in the county of Monmouth and in Herefordshire, where there is splendid land, adapted for the growth of hops, fruit, and well adapted for wheat. In the counties of Hereford,

Worcestershire, and Monmouthshire are very rich red soils, and in South Devon also, in the neighbourhood of Torquay, there are rich soils, and in Cornwall, where there are many districts in which the earliest peas, new potatoes, and vegetables for the London market are reared—all upon the old red sandstone. We meet with them in Cornwall, Devonshire, Monmouth, Herefordshire, Worcestershire, and Brecknockshire; and then again the old red sandstone crops out in Berwickshire to a certain limited extent, but more extensively in Lothian, where it forms the basis of the celebrated East Lothian farming, extending right away down from the Frith of Clyde to Forth, and to Dunbar.

The fertility of the old red sandstone is very high in this part of the United Kingdom, the rents having been to my own knowledge as much as £5 per acre over large areas. I have seen five hundred acres of land lying on the old red sandstone let at £2500 a year; but rents have come down a good deal in late years, though perhaps not so much as some of us would at first imagine. The old red sandstone occupies a position on the east of Scotland—in Caithness, in Cromartie, round the Moray Firth, and it has been pointed out as an extraordinary fact that wheat cultivation can be carried on upon the soils of the old red sandstone further north than on any other formation.

With reference to the remaining formations—the Silurian, the Cambrian, and Laurentian—they run still further west and north. They form the highlands of Devon, of Wales, of the highlands of Cumberland and of Scotland. On these formations are vast areas under sheep or pastoral farming, but we need not expect to find any large amount of land suitable for arable purposes.

CHAPTER VII.

Formations yielding Soils of Clayey, Free, Poor, and Rich Character—Review of Indications of Fertility—Differences between Quality and Value of Land—Sedentary and Transported Soils—Peat Soils—Volcanic Soils—Methods of Improving Soils—Land Drainage—Reasons for its Usefulness.

ALL soils, from whatever geological formation derived, have certain points in common. They all are composed upon the general principle or general arrangement of proximate constituents already named. All contain a certain variable proportion of soluble plant food. All are made up of the four or five familiar substances, sand, clay, lime, and vegetable matter interspersed with mineral fragments. And again, upon every formation good land and bad land, and in very many cases light land and heavy land is to be found.

But, while this is the case, it is equally true that there is a leading character which runs through soils derived from the principal formations, either as regards the general contour of the districts, and consequently of the soils resting upon it, or it may be the fertility or non-fertility of these soils. Certain geological formations may be spoken of as usually characterized by the presence of heavy land, for example the London, the Weald, the gault, the Kimmeridge, the Oxford, and the. lias clays. All these yield soils of stiff character. On the other hand, in certain formations light, easy-working soils predominate, more suitable for sheep-farming in combination with arable cultivation; as, for example, the soils of the upper chalk, of the lower oolite, of the magnesian limestone, and of the Yoredale rocks and the millstone grit.

Again we have formations which have earned the reputation of producing rich soils, and among these may be particularly mentioned alluvial or diluvial soils of mixed origin which accompany the course of rivers, and are especially to be seen around their estuaries, and also forming soils near the coast by the action of tidal currents and tidal deposits. These soils are also of mixed origin, and are therefore sure to be fertile.

Rich soils are frequently to be found where two or more formations meet together, and where the soils are mingled—around the edges of formations. Rich classes of soils are derived from the decay of the lower chalk, the upper greensand, the corn-brash, the lower oolite, the lias clay, the marls of the new red sandstone, the cornstones and marls of the old red sandstone, and the decay of basaltic, trappean, and lava rock. Other formations again yield soils of low average fertility, among which may be mentioned the soils of the upper chalk, of the lower greensand, of the Oxford clay, of the magnesian and mountain limestone, of granite rocks, and of those lower primary stratifications which have been spoken of as the Silurian, Cambrian, and the Laurentian, which frequently rise up into very high mountains. The geological derivation of soils is therefore one of the indications by which we may assist ourselves to come to a conclusion with reference to the quality of soils.

Before leaving the subject we will see how all the previous considerations must be kept steadily in view—the capabilities of the soil as indicated by all sorts of vegetation and growing crops, its texture, its colour, its depth, its sub-soil, its climate, its altitude, its slope or aspect, and its geological position. All these indications ought to assist us in coming to a conclusion as to the quality of land.

I want in the next place to show that when we have arrived at a judgment as to the quality of land we are very far indeed from having arrived at its value. There is a very great difference between judging of the quality of land and of the value of land, either as regards rent or purchasing. Therefore it may be worth while to consider that the quality of land—that is to say, the quality as indicated by the previous numerous considerations—makes one, and only one, of the elements of value. Value depends in a great measure upon position. Fertile land removed from civilization is of no value. Barren land placed in the midst of the city of London is of immense value; so that the value of land is rather fixed by its position than by its fertility. The value of land depends upon the state in which it is found with reference to fencing, roads, cottages, house and buildings, water supply, facilities for working it, its tenure and other considerations, such as the burden of rates or other charges to which the land is liable. It may be in the form of tithe, or it may be in the form of rates of various kinds. All of these matters have to be taken into account before we can in any way fix either the selling or the rental value of land; so we must not confuse quality and value together, and think that because there are certain rules by which to judge of the quality of land, we are in a position to go out and value it.

Soils may be divided into two great classes: first into what are called sedentary soils, which occupy the position in which they were formed, as, for example, when a light-topped chalk soil is found resting upon a chalk foundation, or a strong clay soil resting upon its native lias or its native Oxford clay. Those are sedentary soils. These are the soils which may be fairly presupposed will resemble in general character the rocks upon which they rest. And in the second place we speak of another class of soils which are termed transported soils—soils which have been brought from a distance, and which are best illustrated by a class of soils already named and already described as alluvial. These soils have been sorted and deposited by the action of running water.

In the next place, there is a class of soils which can scarcely be called conformable to either of these. They have neither been derived from rocks, neither have they been transported from a distance, but, like Topsy in *Uncle Tom's Cabin*, they "growed" where they are found; that is to say, peat soils. These peat soils have a vegetable origin, and they usually rest upon a foundation of solid clay, and frequently of gravel. The history of peat soils has been as follows. There has been an interruption of the drainage of the soil owing in many cases to the overblowing of a natural forest. This interruption of the drainage has caused the growth of a certain class of plants known as peat plants, which have gradually accumulated until they have sometimes reached a depth of a hundred feet. In other cases peat bogs are only a few feet in depth. As just stated, they occasionally attain a depth of one hundred feet, and in such cases we must assume that there has been a subsidence of the surface accompanying an upward growth of the peat. The growth of peat upwards in such cases would in some respects resemble that of Darwin's view as to the growth of a coral reef built up from the slowly subsiding bottom of a tropical ocean.

Peat soils occur largely in Ireland, and they occur to a considerable extent in this country, in France, and in Germany, and, in fact, throughout the northern and middle portions of Europe; but in the southern latitude of Europe, and lower latitudes generally, the formation of peat ceases.

Peat soils are rich in organic matter, but they are wanting in mineral constituents. The only mode in which peat soils can be thoroughly developed is by mixing them with the underlying soil, where they make a very useful and even a highly prolific class of land.

The last class of land remaining to be mentioned is that which is derived from volcanic action. It is known to be a very fertile soil, and it is of interest to us as agriculturists, because it not only includes soils recently poured forth from volcanoes during historic times, but also a large area of ground which has been produced by lava streams from extinct volcanoes. There are extensive areas in British India, and even in Scotland, which owe their fertility to the fact that they have been produced by the gradual crumbling down and decay of lava. These soils have been already noticed under the class of basaltic soils.

This must bring to a conclusion the interesting questions of soil formation and soil distribution.

I have, in the next place, to notice certain methods at our disposal for improving soils; these measures are of two descriptions, often spoken of as mechanical and as chemical. The mechanical methods include the improvement of soils by altering their mechanical texture. And there is then left the methods of improving them by the direct addition of fertilizing matter. I consider it more strictly true and a better definition of these two methods to speak of them as means of developing the resources of soils from within themselves; and in the second place, means of adding from without to the wealth of plant food contained in the soil.

With reference to the first class of improvements, land drainage occupies by far the most important position. It has been very properly considered that land drainage lies at the foundation of all land improvement, because until wet land is drained all other expenditures of capital are fruitless. Neither deep tillage, clay burning, liberal treatment, or the feeding of tons and tons of cake and corn can be profitably undertaken. It is necessary, in the first place, that land should be dry; afterwards we may proceed as seems best to carry out other improvements. The history of land drainage is deeply interesting, but it is just one of those sections which may be omitted at present, so that we may proceed to the practical aspects of the question. I cannot pretend to lay before you the whole subject of drainage, as it would itself take up the whole of the present volume.

An important question with reference to drainage is the descriptions of soils that require drainage. We shall probably find that most of the land in this country is already naturally drained. All chalk soils, all those light-topped soils which I have described during the last chapters resting upon fairly porous rocks, are naturally drained, but there are many soils which the water cannot possibly percolate through. They offer a resistance to the downward passage of water, owing to their extremely fine state of division. The pores of the soil, or the inner spaces between the particles of the soil, are so minute that the friction or resistance to the downward passage of water is excessive. That is the case in all the great clay beds, which I need not again enumerate. Also in many local beds of clay in which the particles are so fine, and the inner spaces so minute, that there is a serious obstacle placed in the way of the passage of water. The consequence is that it is arrested at and about the surface, and becomes an evil rather than a blessing. Such soils require artificial drainage.

A second class of soils is wet from *position*. They are naturally capable of allowing the free percolation of water, but at a greater or less depth beneath the surface there is a more or less impervious substratum, and this impervious substratum causes an accumulation of water below the surface soil which gradually forms what is called a "water table," or, as it is sometimes called, a "reservoir." Water is accumulated upon this tenacious and impervious substratum until it reaches the surface and renders it wet. Notice that there is a great difference between the two cases. One is the case of stiff soil, which by its fine condition offers an absolute resistance to the downward passage of water; the other is that of light soils, which are wet simply because at a certain depth beneath the surface there is an impervious substratum which causes an accumulation of water upon it, which accumulation reaches to the surface, or near to the surface, and causes damage to growing plants.

In the case of retentive clay soils, the soils are wet from rainfall. Clay soils are always wet from direct rainfall, not being able to pass off the water sufficiently rapidly, and therefore there is an accumulation of water near the surface.

The second class of soils are often wet from water which falls at a distance, and which finds its way down to lower levels and accumulates there, not from direct rainfall upon the surface, but from the soakage of water from higher levels.

The water economy of these two classes of soils will occupy us shortly, but I want first to ask and to answer another important question, namely, why is it an advantage to rid land of water? We know the benefits which accrue from water, and we know the disasters which are inseparable from a long-continued drought. Every one has heard of the oasis in the desert, which is simply a green spot where there happens to be an abundance of water, and most of us have probably noticed the rank herbage generally associated with plenty of moisture. We are also familiar with the luxuriant growth of the water meadow. Those who have watched a water meadow throughout the season must have been struck with its extraordinary power of growth. Such meadows are mown annually. There is a heavy crop of hay taken off them every year, and that has been going on from time immemorial. And not only do they give a crop of hay every year without any manure, but they likewise give abundance of spring keep for sheep, and a plentiful autumn feed for horned stock and horses, all being brought about by a copious and constant supply of water.

Hence the question as to why it should be such an advantage to take water out of land becomes very important.

In the first place, I should say that by drainage we get rid of *stagnant* water. That is the great advantage in drainage. It is not that we get rid of water, but that we get rid of stagnant or effete water. Stagnant water is exhausted of its oxygen by contact with decaying vegetable matter. It has therefore become soured. In the next place, it keeps out fresh supplies of water. It is leading a sort of dog-in-the-manger existence—useless itself, effete itself, and at the same time preventing the accession of fresh water.

Now if we can rid land of stagnant water, if we can so manage as to allow a free passage of water through the soil, we make way for fresh water, and that fresh water is thoroughly ærated. It likewise contains those stores of nourishment which rain-water brings down from the atmosphere. It contains a certain proportion of carbonic acid gas. It contains nitrogen, it contains ammonia, and also a very considerable portion of chlorine in the form of common salt. This water is brought into close contact with the growing roots, and then passes onward. The fact that the water is in a state of movement—that it ceases to be stagnant—is a perfect guarantee that it is accompanied with and followed by air, so that the soil becomes oxidized. The iron salts in the soil are maintained in a perfect state of oxidation. The vegetable products in the soil are converted into available plant food. All the numerous benefits which a free accession of air into the soil can give are in this way obtained.

In the next place drainage has produced an alternation of condition from wet to dry and from dry to wet; whereas an undrained soil is always in a water-logged condition. The drained soil is alternately wet and dry, while in the undrained condition it is in a constant state of wetness. The consequence of these alternations is pulverization of the soil, and this pulverization is carried to a very considerable depth, so that clay soils which have been subjected to the operation of drainage gradually deepen and mellow to such a degree that the original character of the soil and even of the sub-soil becomes completely altered.

Besides, we must not forget the improvement in the climate of a drained soil. The average temperature of such a soil is considerably enhanced, due in a great measure to the checking of evaporation. A wet soil, like a wet garment, is cold. It is constantly chilling the roots of the plants and the body of the soil. Not only is the cooling of the soil due to evaporation, but it is also owing to the peculiar mode of cooling of water and other fluids when exposed to a low temperature. During the winter season, when wet land is exposed to cold air, and when it radiates its heat out into a clear sky, the upper film of water becomes cold, and as this film increases in weight or in density it sinks into the lower layers of the water-logged soil. Its place is taken by warmer water from below, and thus there is a gradual cooling of the entire body of the water within the soil to a very great depth. Notice that this is different from evaporation. Evaporation no doubt chills the soil, but the soil is chilled to a very considerable depth owing to "convection," which is the law of the cooling of fluids. Similarly the wet soil is never able to obtain the benefits of summer heat, because if a thoroughly wet soil is exposed to the sun's heat, the upper film of water becomes warm, and remains at the surface or is evaporated from the surface, but being the lightest, the lower layers of water cannot rise through it. The temperature of the bottom water remains at a low degree, the entire solar heat being consumed in the evaporation of the surface water.

These appear to me to be the chief reasons why drainage is beneficial. This is a question which I have frequently asked in my capacity of examiner, and the answers which I receive are very often too much restricted to one or two considerations bearing upon plant food in the soil. I am told generally that the advantage of draining is that it causes the passage from the dormant condition of soil into the active condition. I have no doubt that it does. We might add this to the previous considerations. The free entrance of air, the free entrance of water charged with carbonic acid gas, the constant contraction and expansion of the soil, all help to reduce soils and to reduce the mineral matter of soils from an insoluble and unavailable condition into an active and available condition. But there are a great many more considerations, and those considerations we ought to see brought out in the answers which are given to such a question as this.

Unfortunately this change from a dormant condition of soil to an active condition is used as an answer to such a vast variety of questions, that when we read so much about these dormant constituents and their change into an active state, we feel inclined to become a little impatient. For instance, if we ask in a question-paper Why farmyard manure is such an excellent manure? we are told, "Because it changes the dormant constituents into active constituents." We may ask Why autumn cultivation is excellent? and the answer is, "Because it changes the dormant constituents into active constituents." Or we may ask for the theory of tillage operations, and the theory of all tillage operations appears to be, "that they change the dormant constituents into active constituents." Or we may ask a drainage question, and are told the same thing.

Now of course we want, if possible, to take both teachers and students out of anything like a narrow groove, and it is for that reason that I think all the other features ought to be brought prominently before students, and they should be encouraged to look upon the subject in a wider manner, and from as great a number of points of view as possible.

CHAPTER VIII.

Beneficial Effects of Vegetation on Soils after Drainage—Action of Drains in Light Soils—Action of Drains on Retentive Soils—Smith's System—Elkington's System—Advantages of Drainage—Improvement of Land by Trench Ploughing—Sub-soiling—Clay-Burning—Claying—Marling—Chalking—Warping—Ordinary Cultivation.

WITHOUT reverting to the various reasons why land drainage is beneficial, there is one important point which has been omitted, namely, the action of vegetation, which no doubt tends to modify soils and to improve them after they have been freed from surplus water. The free ingress of the air has been mentioned, and the oxidizing effect which it exerts not only upon sour matters in the soil, but especially on sour matters in the sub-soil; the effect being, as was pointed out, pulverization of the sub-soil as well as of the soil, and that down as far as the depth of the drains. So that if we imagine the case of a thoroughly stiff soil resting upon a thoroughly stiff sub-soil, the tendency after thorough drainage is to modify, pulverize, and oxidize the whole of the section of sub-soil above the drain, and to cause it to assume in some degree the character of a free soil. Now it is just at this point that the action of vegetation comes in; the roots have a freer ingress as well as the air; the altered and improved condition of the soil increases the depth available for plants, and we have not only pulverization taking place, but the permeation of the ground by vegetable roots, which as they decay gradually alter the nature of the sub-soil. There is at Rothamstead a very interesting series, exhibited in glass tubes of eighteen or twenty inches in diameter. Each of these wide tubes contains a section of soil taken from the surface to about three feet in depth, and there may be seen the wonderful effect produced by drainage and by good surface cultivation and liberal treatment in gradually altering and modifying the sub-soil—altering it from the condition of

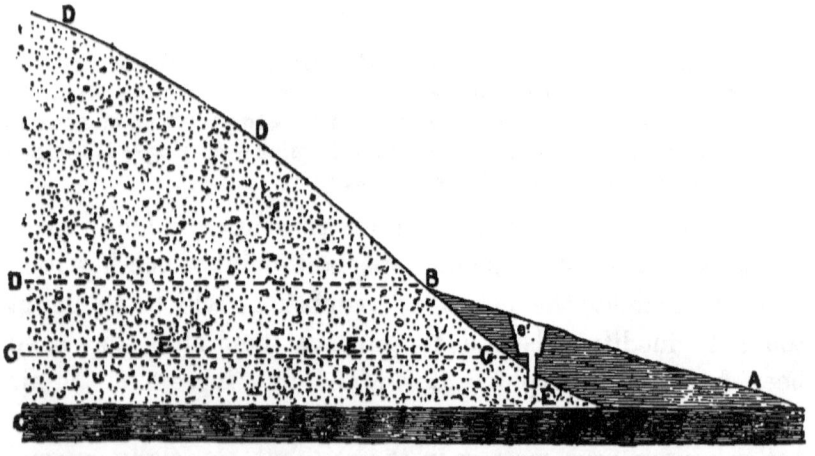

FIG. 2.—Portion A B wet from accumulations of collected water below it which rises up through it. E E E super-saturated portion of underlying soil D B the water table. A springs which overflow and help to wet the surface and injure vegetation. D D dry porous soil allowing of the free percolation of water. This diagram also illustrates Elkington's system of drainage.

a blue clay of a compact soapy texture, and converting it into a red crumbly sub-soil which must be better in all respects for the growth and development of crops.

Taking up the subject where we left it, I pointed out in the last chapter the difference between heavy lands and light lands, either of which may be incommoded with surplus water. Heavy soil is wet from the direct resistance it offers to the passage of water through its mass. Light soils are wet from the obstruction of a clay bed or other impervious bed beneath the surface. If there is no such obstruction to the downward passage of water, such lands cannot be wet; but if such lands are wet at all it must be from the presence, at some distance beneath the surface, more or less, of an obstructive substratum. The presence of an obstructive substratum may therefore in such a case be assumed. In some cases this porous soil may be wet from rainfall upon its own surface, but in more ordinary circumstances it is also wet from the soakage of soils from higher lands. This is shown by the accompanying diagram. The portion marked A B may be considered to extend over several acres. At D there is a porous soil upon which water falls and passes away, and so far as the summit of the hill is concerned it is completely dry. The water, however, passes downwards until it meets with the obstruction at C C, and there it collects. Here we have the case of a retentive soil, resting at the tail of a gravel bed, and such a soil becomes wet not only from the downward passage of water from direct rainfall upon it, but likewise from bottom water. Water rises upwards as well as sinks downwards. Here is a case in which a porous soil is not only wet from direct rainfall, but from position. The first class of soils, that is the clay soils, may be said to be wet from top to bottom. If we take the surface of a clay field, the twenty-six or thirty inches of rain which annually falls upon it is stated by our great authority, Sir John Lawes, to be disposed of in the following manner—seventy to seventy-five per cent. is evaporated, or exhaled from the leaves of growing crops, and the remaining twenty-five or sometimes thirty per cent. percolates. Therefore in the rainfall of thirty inches we may readily calculate how much evaporates into the air and how much percolates, or is disposed of in other ways.

Impervious soils must become wet on the top. It is the upper portion which will become puddled and wet with rainfall to its great injury, and the water instead of passing through the soil runs over the surface, and carries with it not only fertilizing matter, but the finer particles of soil, and deposits those finer particles in the open furrows, or at the bottom of the slopes. Such soils are wet from top to bottom, and if a trench is dug the ground is found comparatively dry, and pipe-tiles would not convey any quantity of water from such a soil in its then condition.

I may quote a case which happened to my own father when draining his land in the north of England. His friends found fault with his method, and asked why he was draining land, when the dry condition of the newly-cut trenches gave very little indication that there was any overplus of water. His reply was, "I drain the ground to get water into the land as well as to get it out of the land." Such lands, as we have already seen, offer a tremendous resistance to the passage of water, and the action of the drain is in fact at first very gradual. Drains upon such soils act slowly at first, until, by their pulverizing action, they gradually transform the soil into a more porous condition, and the amount of work done by the drains gradually increases from year to year. The pulverizing action begins close to the drains, but it gradually spreads outwards on both sides and especially above them; it radiates from them until it meets the action of the contiguous drains, which ought not to be at too great a distance, let us say not more than twenty-one feet, in which case ten feet six inches is all the distance that the drain is required to act upon until it meets the action of the adjacent drains. Putting these drains in at twenty-one feet apart, the pulverizing action of the drain proceeds from one drain to the other until it meets, and that no doubt is the proper explanation of what is called the reciprocal action of drains, which is a well-known term with reference to the action of drains in clay lands. The idea of a reciprocal action of drains has arisen from the fact that they appear to assist each other; and also because a single or solitary drain in a clay field exerts only a small effect. A solitary drain does not act, because the pulverizing action which it exerts is speedily lost in the mass of clay. It may act for a few feet on either side of it, and the effect then ceases, whereas if we make a series of drains, it may be twenty-one feet, or in some cases only fifteen or sixteen feet apart, then each of those drains, acting from its own centre, gradually pulverizes the soil. The influence of the drains meets at the centre, and we get that thorough pulverizing through the entire clay bed which ends at last in all the advantages of thorough drainage. It is clear then in this case that the nature and character of the soil greatly affects the system of drainage which is employed. Clay lands—by which we mean both clay soil and subsoil—evidently require a thorough or a regular system of drainage, or what has been called the furrow system. It requires that the ground shall be regularly and equally divided into panes of not too great width. This is Smith's system, called after Mr. Wm. Smith of Deanston in the county of Perth, who gave his system to the world in 1823. It is the regular, or through system of drainage, and is the best for stiff clay lands.

Passing to the consideration of the other class of soils, they may be properly said to be wet from *bottom to top*. The water passes down through such porous soil; for rain will always pass by the law of gravity towards the centre of the earth in a straight line, or as straight a line as is possible, and if there is no obstruction it will continue to go for hundreds of feet, until in fact it is stopped. That is why we have deep wells as well as shallow wells. In the case of deep wells the water has met no special obstruction, and it goes on until it meets one of those deep clay beds which are sure to occur somewhere in the vertical section. Now if a soil is wet to such a degree as to interfere with growing crops, there is probably a clay bed some few feet beneath the surface; the water sinks until it can sink no further, and it then begins to accumulate on the clay bed, and rises upwards towards the surface. Possibly it may rise above the surface, in which case we have a mere or marsh. Sometimes it rises to within a few inches of the surface, in which case an un-

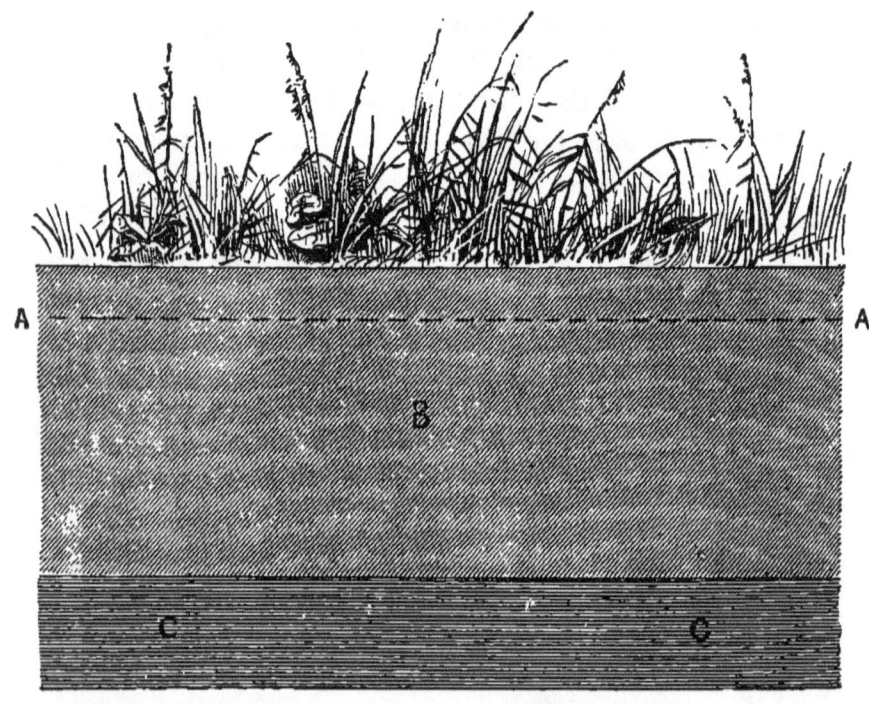

FIG. 3.—SHOWING SECTION WET FROM AN IMPERVIOUS SUBSTRATUM.

A A the upward limit of water, or water table, causing a rank and coarse vegetation. B the section water-logged, or super-saturated with water. C C retentive clay bed preventing the escape of water.

wholesome condition of soil is induced. At other times it lies well beneath the surface, in which case we should have a wholesome condition of soil. We have in these light soils a section which has been properly called the section of super-saturation, above the obstructive bed, or a section which is thoroughly water-logged, the section of super-saturation, sometimes called the reservoir, which reservoir is bounded on its upper surface by what is called the "water-table." The dotted line A A on the above diagram shows the water-table. The upper margin or surface of the super-saturated portion is called the water-table. Above the water-table succeeds a section of soil which is wet from capillary attraction. This I have shown by direct experiment varies in thickness according to the nature of the soil. Capillarity in sand does not appear to raise water more than about sixteen inches, and therefore for sixteen inches above the water-table we may expect that in a sandy soil the section will be wet from capillarity. In a very finely-divided clay capillarity appears to be able to raise water about thirty-six inches; fine clay soils will therefore raise water by capillarity about a yard. Above the portion which is wet from capillary attraction we have dry soil, unless wet from recent rains. Three zones have been mentioned, the super-saturated portion, or reservoir; secondly, the portion beyond that which is wet from capillarity; and thirdly, a higher portion which is dry. If this dry portion does not exist, that is to say, if the water-table is so near the surface that capillarity extends to the surface, several evil effects will be produced. The solid saline portions of the water are left on the surface. These form incrustations which in some cases take place to such a degree as to become a positive bar to cultivation.

Also, when this capillary action extends up to the surface a constant chilling of the soil takes place. The water-table ought to be lowered artificially to such a depth that capillary action cannot extend to within one foot of the surface, so that we may have a dry blanket, if we may so speak, upon the surface, and that the roots of plants may be able to dip down easily into the portion which is wet from capillary attraction, or even to the reservoir if it suits them; but the top portion must be dry. When the water-table is reduced to such a distance beneath the surface that we have this improved condition of things, the water will do no injury whatever to farming operations.

If the soil is wet and requires drainage, of course we have to bring about this desirable state of things, and it is done by the placing of drains at sufficient intervals apart. Now what happens if we place a drain in a porous soil, say, four feet beneath the surface, and three feet below the level of the water-table? That drain begins to run just as surely as if we were to knock a bung out of a full cask. The land being free, the air easily follows, and there is not that resistance to the passage of water which clay land offers, so that the distance between these drains may sometimes be greatly increased. Where there is a thoroughly porous or gravelly soil wet from position in the way I have explained, it is extraordinary to what distances drains will act. I know a case of a farm in Gloucestershire which was entirely drained without the use of a single pipe-tile. Merely by cleaning out and deepening the ditches which intersected the fields such a complete circulation of water was induced that the whole intermediate ground was dryed—the water-table was lowered. It is not uncommon to find surface-wells dried by neighbouring drainage works, and in such cases the conditions are similar to what I have described. There has been an underlying layer of gravel or open material, and the moment the water was allowed free egress it flowed out rapidly; and the action of these drains extends a long way on either side.

This leads me to speak of the rival system of drainage known as Elkington's system. It is much older than Smith's, and was first practised in the year 1763 by Mr. Elkington of Princethorpe, Warwickshire. Mr. Elkington was engaged in the work of draining his farm when he met with the following experience. The ground was wet, the trench which he had dug did not appear to be sufficient. Then we have the story of the shepherd with his bar over his shoulders. Mr. Elkington asked for the shepherd's bar, and plunged it down through the bottom of the drain, and up came the water, almost giving the idea of the miraculous. Mr. Elkington was considered to be almost a wizard in his time, and to possess the power of divining water. I am not sure how far he encouraged that idea, but he was certainly credited with the power of being able to use the divining-rod, and of knowing exactly where to look for sources of water. Mr. Elkington plunged the shepherd's bar down through the drain, and up came the water. He afterwards used an augur contrived especially for this purpose, most successfully on his own farm. He was employed in Warwickshire very largely, and subsequently in all parts of England, and finally in 1795 Parliament granted him £1000 for his services to his country. Mr. Elkington's system of drainage could not be carried out everywhere; we might in vain strike the shepherd's bar into a hard clay, but a state of things illustrating his system is depicted in diagram 2, p. 98. Springs pour out at B, and run over the surface, and the water pressure from below is constantly pushing up from B C. Mr. Elkington's trench was therefore perfectly futile, but when he tapped the tail of the water-bearing stratum, as shown on the diagram, up came the water with force, until at length the level of the water D B became transferred to the level of G G, and the whole of this ground was relieved from the upward water pressure, and the springs being tapped, all this land, possibly several acres in extent, would be drained. The ground is wet with water pressing upwards, the trench being made to communicate with the sand or gravel which is below, and at once a tendency is induced to lower the water-table to the requisite depth. Now it may occur to us that such circumstances are not very common. Mr. Elkington, living as he did in Warwickshire, where the lower oolites, the lias clay, and the new red sandstone all verge upon each other, they furnished the peculiar alternation of retentive and porous strata which are necessary; but it is scarcely likely that we shall find many localities where land can be drained according to this dexterous method. Still, clay soils which are underlaid by a gravel bed may in some measure be treated on Elkington's principle. For example, we may take the case of a clay soil where some six feet beneath the surface occurs a porous gravel bed. To drain such ground as that three feet deep would be a mistake. It would be much better to take a leaf out of Elkington's book, and place the drains seven feet deep, get them down well into the gravel bed, and then we should have a sort of connective medium between the drains which would enable us to place them at great distances apart. Instead of placing our drains in the compact clay three feet deep and twenty-one feet apart, we should sink them seven feet in depth, and with the aid of the gravelly bed which lies under the clay, increase our intervals to forty or fifty yards apart! This is well known to practical drainers, who sometimes go as deep as ten or eleven feet in order to tap a porous substratum.

Before draining a large tract a thorough examination of the soil section should be made. Sometimes drains are constructed partly after Elkington's system and partly after Smith's system. Springs must be tapped, and hidden shifting sands must be reached by a separate system of drains placed below the contemplated depth of the ordinary regular drains. Then having relieved the land from its springs we proceed with our regular system of drains eighteen, twenty, or twenty-five feet apart, the traversing drains being laid a few inches deeper. Again, sometimes on hill-sides it is necessary to cut off the soakage water from higher levels by a drain cut across the slope, and then proceed to drain the surface below on Smith's regular system.

I wish to show in this little work that Agriculture is a many-sided subject, and that numbers of considerations must be taken into account in looking at any agricultural problem. The practical advantages of land drainage are many. If a farmer pure and simple were asked why he approved of drainage he would not reply in the terms in which I explained their advantages in the previous chapter. He would not tell us he drained land because it increased the underground temperature or checked evaporation; but he would tell us he drained because he got better crops. That is one of the effects produced by drainage, not only better crops, but a better quality—better barley and better wheat, and less mildew and less blight. And we also get a much larger variety of crops. We are no longer restricted to beans and wheat on wet clay lands, but we can introduce turnips, swedes, mangels, kohl-rabi, rape, clover, and many other crops. We get a much cheaper tillage, and the number of our working days in the year on clay lands is much increased. One of the difficulties farmers on clay lands have to deal with is that the number of days in the year on which they can work the land is very limited. Light land farmers can always work their land as long as it is dry overhead, but a clay land farmer cannot do that. He must keep off his land for days or even weeks after heavy rains; he knows that the carters would be better employed toasting their toes over the fire than ploughing the ground in wet weather. The consequence is that the clay land farmer must keep a force of horses for emergencies. Now by drainage we lessen that evil—the number of working days is very much increased, and therefore fewer horses need be kept, or those horses can be kept in good condition on cheaper food. The health of our live-stock is also improved by the drainage of the ground—

especially the pasture-ground. Certain diseases disappear with drainage. Redwater, for instance, a most destructive scourge, frequently to be found on undrained land, disappears when the ground is dry. So also with liver-fluke or rot in sheep; and black quarter or blackleg, that scourge of stockbreeders, which often takes off the very best of their young stock. And not only so, but the human population is healthier. Ague and consumption disappear after land has been drained, and it is not among the least of the advantages of a thorough system of drainage that the health of the population is much improved. Some people cry out against Government assistance to landlords in the form of drainage grants; they consider that it is wrong for the Government to help landed proprietors with public money; they think that landed proprietors should help themselves. But drainage is a philanthropic work, because among other great advantages the health of the population and the whole interests of the country benefit by the drainage of land, and if a landlord is short of capital, and can borrow from the Government on the principle of repayment with interest in twenty-five or thirty years, it is an excellent thing for the country.

Next, with reference to other methods of land improvement. After we have drained our wet land we may proceed to lay out our money with some prospect of getting it back again. Deepening the cultivated soil is the next thing, and that leads me to the consideration of subsoiling, trench-ploughing, and deep cultivation. Trench-ploughing or double ploughing is an agricultural operation which must be conducted with very great care. It is one of those thorough-paced improvements which finds favour with the Press, but is often a way to bury money. I will tell you what it often, but not always, does; it brings up an intractable, unweathered, sour subsoil, and the consequence is it dilutes the soil and lowers its percentage of available fertilizing materials. Secondly, it sometimes brings up hidden matter of sour and injurious character which did no harm as long as it was below the plough-sole, but which does harm mixed with the surface soil—sour unoxidized products, which have been known not only to stop all crops growing, but even all weeds, and to render the ground for a time absolutely barren.

It will also sometimes bring up a most extraordinary crop of weeds, and that has given rise in the minds of many farmers to a firm belief in spontaneous generation. Farmers will tell you that if they plough deep they will fill their fields with charlock, and so they will. If they plough even a little deeper than they ought they will stock their ground with such a crop of charlock as it may take years to get rid of. Such an extraordinary fact requires a few words of explanation. If we walk over the fields in the height of summer we will see how they are cracked, a walking-stick can sometimes be pushed down the cracks a yard deep, and how much deeper they extend it would be difficult to say. Charlock has a small, round, compact and oleaginous seed, and it no doubt trickles down these cracks probably two and even three feet beneath the surface, and when the ground is ploughed to a greater depth than usual we bring up the charlock and reap the evil consequences of so doing.

These are some of the disadvantages of too deep ploughing, but in some cases trench-ploughing is useful. It must be considered good when we have a continuation of good soil to a considerable depth, as in alluvial soils or riverside lands, where a soil may be two or three feet deep. In such cases we shall probably produce a good effect. On loamy soils through which the air has free access we need not fear the presence of poisonous materials; they have long ago been oxidized and neutralized.

The best method, perhaps, is to deepen the soil a little every time it comes into course for roots, or about once in four or it may be five years. The extra depth of furrow should be given in the autumn or winter, so as to allow the frost to weather the newly brought up soil, and in this way the depth of the soil will be gradually deepened without injury. I have a great horror of heroic methods with reference to agriculture; they bring discredit on science, and they bring ruin on the people that adopt them. There is nothing so dangerous as when a man gets hold of notions and begins to figure as a scientific agriculturist in his neighbourhood. He is sure to bring ridicule on science, and injury and loss to himself. We must remember that "various motives fire the strife," and that journalists and agricultural writers of the day, implement makers, and seedsmen are all alive to the importance of publicity, and push certain things before us which great Agricultural Societies are compelled to recognize in some measure or incur the imputation of being "behind the times." Consequently we get a great many things introduced to our notice and paraded in the newspapers which must be guarded against carefully, and only allowed to filter very gradually indeed into our farming practice. Of one thing we may be sure, that it is the most difficult thing in the world to introduce new methods into an old business. We have heard of putting new cloth into an old garment, and it is just as difficult to engraft new notions upon an old-established pursuit like agriculture—it is exceedingly difficult. I could enlarge upon how few of the new-fangled ideas which I have seen brought out during the last thirty years have been adopted and really taken a place in our system of agriculture. They serve their purpose; they often do more good to the men who propose them than to anybody else.

The foregoing remarks arose out of my suggestion that we should gradually deepen our soils instead of resorting to trench-ploughing.

Subsoil ploughing is different to trench-ploughing. It is not the bringing up of the subsoil, but the smashing of it only. Sometimes it is called "knifing," when a deep coulter is fixed upon a steam-plough and dragged through the subsoil, cutting it at intervals of every few inches. Subsoiling does not contemplate mixing, so that many of the evils of trench-ploughing are avoided in subsoil ploughing. It may be defined as breaking the subsoil without bringing it to the surface. Subsoiling facilitates drainage; it also does away with " pans " of various kinds, whether ochreous pans, gravelly indurated pans which have resulted from the constant traffic of horses year after year upon some five or six inches depth of soil, calcareous pans, or moorland pans. Wherever we have one of these pans—which prevents the water passing through—there we have a proper field for subsoil ploughing. Moorband pans and other forms of indurated soil are well known in Scotland. They occur a few inches beneath the surface, on unreclaimed or newly reclaimed soil, and when once broken up they do not form again. Therefore in such cases subsoiling is of very great use. Subsoiling is generally considered to be most beneficial on light, gravelly, chalky, or rubbly subsoils, but is never considered to be beneficial on heavy soils because the clay very quickly closes up again, and relapses into its former condition, the ground is as plastic as ever, and the outlay is wasted.

Clay-burning is a beneficial operation for clay lands. No one was a greater advocate of this process than the late Mr. J. J. Mechi. He was a very clever man, mistaken in some things, but he was not ruined by farming. People sometimes tell us that Mr. Mechi was ruined by farming, but he was not; he was a heavy loser by a Bank. Another very excellent man whose name is associated with clay-burning was the late Mr. Charles Randall, who managed both for himself and the Duke d'Aumale. He greatly improved his farms near

Evesham, on the lias clay, by many operations, amongst others clay-burning. The late Philip Pusey, one of the greatest agriculturalists that ever lived, was also an advocate of clay-burning. Clay-burning is literally the burning of the staple of the soil; it is not mere weed-burning, it is not paring and burning. The ground is broken up during the summer, with a strong three or four-horse plough. The clay is then burnt in heaps placed at intervals on the surface. The fire is first kindled by brushwood, roots of trees, and coal where it can be obtained cheaply. The clay is piled upon the fire in large clamps and burnt, and the burnt clay is then spread. Great pains must be taken not to overburn the clay; it must not be burnt as if intended for a railway embankment or a road foundation—nor into clinking, bricklike ashes, but into a black ash or brown clay-coloured ash which will fall readily to powder when rain falls upon it. The clay is much mitigated by the process. Instead of a plastic mass, the particles become repellent to each other, and we introduce sixty to one hundred cubic yards per acre of a loose material which greatly improves the mechanical condition of a clay field. If not overburnt, the soluble matter in the clay is increased, that is to say, the iron, soda, and potash. The amount of soluble phosphoric acid is slightly reduced. One thing we must not forget, that we lose nitrogen and organic matter. These are drawbacks, but on the whole the mechanical or textural improvement, and the increase of soluble potash and iron, account for the favourable results which follow the process.

We come in the next place to what is called claying; an operation which may well be grouped with marling, chalking, and the mixing of soils generally. All of these similar operations may be considered together. Sometimes clay, sometimes chalk, and sometimes marl or clay are dug and spread on the surface, and if we have a tract of shifting sand it is worth consideration whether by excavating into the soil we can get such materials. Thirty or forty yards of clay per acre has a wonderful effect in binding together sands; but a clay soil will swallow up a large amount of sand without any effect being produced. In some counties, as in Norfolk for example, marling is thoroughly appreciated; marl-pits are opened, and forty to sixty cartloads are applied per acre. The lower chalk is the best material for chalking. Large quantities of chalk are brought up the Thames by barge and applied to the heavy soils of the London clay, and to the heavy marine clays which extend still nearer to the Essex coast.

The mixing and moving of soils may sometimes be carried out with benefit. Where clay and sandy soils exist in close proximity an interchange may be effected with good results, although the effect of clay upon sandy soils will always be more apparent than the reverse operation of adding sand to clay.

The carting of soil on to bare brows is a work that may profitably occupy horses during winter. The tendency for soils to slip down hill-sides and gradually accumulate at the bottom of slopes is well known. Every tillage operation tends to effect this result, and in process of time the upper portion of the hill becomes denuded of soil, while there may be an excess of earth at the bottom. The restoration of this soil to the upper portion of the field is a beneficial act.

Warping has always been recognized as a method of not only improving, but absolutely making land. The area over which warping may be practised in this country is restricted to the neighbourhood of certain rivers in Yorkshire. The Humber, Ouse, and Trent bring down mud in large quantities and deposit it naturally on either side of the estuary of the Humber, thereby forming those alluvial soils on both banks which have already received notice. What is called warping is simply the artificial regulation of this natural deposition of sediment. Tidal action is a necessary adjunct, as the rise and fall of the tide, especially in spring tides, enables the warper to alternately flood and free the area upon which he is operating. The water from the river is allowed to enter artificial canals at high tide by means of sluices, and is conducted to the "compartment" which is to be covered with warp. The water is then allowed to spread itself over the area, the expansion being regulated in order to secure a uniform deposit, by means of what are known as "call-banks." At the fall of the tide the sluices are opened and the water is allowed to flow back into the river channel. Successive tides are employed for repeating the process, until a regular deposit of soil has been formed to a depth of from one to three feet. The soil thus made is of exceptionally rich character. As a case of natural warping the valley of the Nile may be mentioned. The remarkable fertility of Egypt has been produced and kept up for thousands of years by the annual inundation of the Nile, and the consequent deposition of mud brought from Central Africa and the districts near and more remote which it washes in its course towards the Delta.

Another means of improving the physical character of soils is good general cultivation. Thorough good ploughing and seasonable cultivation of all sorts no doubt improves land; it accelerates the disintegration of mineral matter which is constantly going on in soils. The soil is readily exhausted of its available plant-food by over-cropping, but the natural forces will continue to act, and the land will recoup itself out of the vast store of insoluble materials which it contains. This natural process is accelerated by cultivation; the ground is stirred, fresh surfaces are exposed, it is divided and redivided, and there is a more complete exposure to the forces which convert the insoluble or dormant materials into an active state. More than that; there is a thorough aeration of the soil, a point of great importance not only on account of its sweetening effect on the soil, but from its assistance in the important process of nitrification or the oxidation of nitrogen under the action of bacteria. The promotion of this process is now one of the best recognized uses of tillage. Tillage also has a third effect, it cleanses the ground of weeds, a function but rarely mentioned to us by students of the Science and Art Department. Surely from an agriculturist's point of view it is almost the chief reason for tillage; but how seldom do we, as examiners, find it even alluded to.

A good seed-bed must be clean, fine, moist, and rich. If it can be made to answer this description it is a good seed-bed, and the tillage has been thorough. Cultivation perhaps shows the skill of the farmer more than any other part of his business. This statement may not be true on all classes of land, because there are some soils very easy to manage; but where we have clay soils to deal with, the success of the farmer depends more on the way he times his tillages than upon anything else. It is therefore of great importance that he should understand the times and seasons at which it is best to till his ground. It is impossible to communicate such knowledge. It can only be arrived at by experience. A few rules may however be useful as guides to those who have or may have the management of clay soils committed to them. One of the first is never to touch it when it is wet. In the next place, autumn cultivation particularly appeals to the clay-land farmer, because if ever clay land approaches to the character of light land in friability and looseness it is during the period that succeeds harvest and precedes winter. If ever we are likely to be deceived as to the mellowness or friable nature of a clay soil it is at this period of the year. Clay lands work mellowly at that time, and then is the season to work and to clean them. The weeds are weak after harvest, their roots run near the surface, the ground can be easily pared, and in the fine September and October days we can burn the weeds. This is also the time to plough clay lands; the ground being thus exposed to the effect of the winter's frost. One of the finest unpatented tillage implements is frost, and especially is it found useful on heavy classes of clay lands. It gives a pulverized surface much finer than anything man can produce. It is a tilth that will not run together. If a tilth is forced with rollers and clod-crushers, the first heavy rain will cause it to run together like mortar; but a natural tilth produced by the severance of the particles by frost will not run together; it keeps loose. Therefore if we wish to grow roots on clay land we must begin operating in the autumn. Clean it in the autumn, because that is the best time to clean it; and dung it in the autumn if practicable. Clay land is all the better for plenty of long strawy dung; every straw rots and leaves a space which facilitates drainage and opens up the clay. Clay soils also possess exceptional powers for retaining fertilizing matter. They will not allow the valuable ingredients to wash through them, so that we may safely dung them in the autumn. We then give it a deep ploughing, and the effect of such treatment will be a fine tilth in early spring. If we can drill our rape or our mangolds upon the fine loose mould which is produced by the winter frost, we secure the conditions for the successful growth of roots.

Another plan for dealing with such soils is to raise them up by means of a double mould-board-plough into ridges twenty-five to twenty-seven inches apart from crown to crown, place dung in the bottom of the trenches, and split the ridges over the dung, leaving what was the centre of the ridge as the hollow. In the spring the ridges will be found to have mellowed down—the frost has gone right through them. Then harrow the ridges in the line of their length with a curved harrow made for the purpose to catch both centre and sides of the ridge. Next use the double mould-board-plough so as to form up the ridges again, and we shall have a beautiful seed-bed for mangold-wurzel or for swedes, which we may at once proceed to drill. In this way we can get clay lands brought into the requisite state of fineness in the month of April or May for the reception of mangel, rape, or turnip-seed.

CHAPTER IX.

Ordinary Cultivation (*continued*)—Main Differences between the Cultivation of Stiff and Light Soils—Root Cultivation—Autumn Cultivation—How to clean a Foul Field.

THE subject of land cultivation is approachable from several points of view. In the first place, as it varies with the character of the soil; secondly, as it changes with the crop for which it is intended; and thirdly, as it is performed by numerous implements, each of which may be separately studied—as, for example, by the plough, by the harrow, by the roller, by steam cultivation or horse cultivation. Each of these aspects comprises a great deal that is interesting and useful, and we cannot do better than follow out the lines indicated.

In the first place, then, we shall take the main differences between the cultivation of stiff land and light land, which after all are the two principal classes of soils. We generally speak of soils either as stiff corn lands or easy working free lands, adapted for turnip husbandry and sheep. With reference, then, to the cultivation of stiff land, I have already laid stress on the importance of tilling it when it is in a fit and proper condition—that is, when it is dry; we have also noticed the advisability of autumn cultivation. Both these points are of equal importance in the cultivation of stiff land. It was also recommended that stiff grounds should be treated with farmyard manure in its bulkiest condition. I am not going to recommend manure-heaps and the treatment of manure-heaps in connection with the farming of stiff land.

I am going to recommend that the manure produced in the yards should not be treated except by taking it fresh from the place where it was made, in what is called the long, green, or fresh state, spreading it upon the surface, and ploughing it in, so that any fermentation or any alteration in its bulk may take place within the recesses of the soil, thereby assisting to break down and pulverize the ground, to liberate its carbonic acid and other gaseous matter, and to add to the bulk of organic matter within it, and thus to mitigate and improve its quality.

With further reference to the principles involved in the tillages of clay land, as far as possible the tillages upon clay land ought to be done in harmony with atmospheric changes. We must work in harmony with the forces of nature, and certainly not contrary to them. Anything like a forced tilth or an effort to grind or grate down the soil into a powdery condition is to be avoided. There is no greater mistake than to endeavour to pulverize clay lands by the use of heavy Crosskill clod-crushers, Norwegian harrows, or other implements designed to cut and pulverize clods. You cannot force a tilth without vast expense: without sacrificing moisture, or producing a condition of soil too dry for the germination of seed, and without consolidating the under section to an injurious extent. While you may obtain perforce a fine surface, you are very liable to produce a stiff, leathery condition of soil within three or four inches of the surface— that is, you lose depth of tilth.

All of these evil consequences are to be avoided by timing tillages judiciously. Hence once more we are brought face to face with the advantage of autumn cultivation, and the utility of frost, changes of temperature, and changes of condition and of temperature during the winter months. Equally true is it that we may use the changes of temperature in the summer as well as in the winter, a fact frequently overlooked. We hear of the pulverizing effect of frost, but we hear very little of the pulverizing effect of the solar rays, and of the alternations between a moist and a dry condition, which assuredly will break down tenacious clays even in the summer. A judicious waiting for rain is a guiding principle to the clay land farmer. He ploughs up his stiff clay land, and there lets it "make" or pulverize under the combined action of the changing temperature between night and day, the scorching effect of the sun, and the ameliorating and mellowing influence of thunderstorms, or of sharp showers of rain. In working a bare fallow, for example, which is a system of cultivation still in vogue on certain classes of clay land, the object is to produce the roughest possible condition of soil at midsummer—that is the way to produce an efficient bare fallow; the ground being then strewn over with large clods, sometimes described as "nags' heads." These great clods are rumbled and tumbled about by heavy drags, or it may be by other toothed cultivating implements, and during the period between Midsummer Day and the 1st of August they crumble down under the influence of tillage implements, assisted by showers, thunderstorms, and these invaluable changes of temperature and pulverizing influences which are constantly assisting the farmer if he will only allow them; so that, by the end of August, we find instead of a surface covered with "nags' heads," we have a tilth composed partly of loose material calculated to assist in the germination of seed, interspersed with moderate-sized clods, and that I take to be the most suitable condition of clay soil for wheat-sowing, a condition in which there is a sufficient amount of fine soil, and intermixed with it, and upon the surface of it, a nice round clod.

When the wheat is drilled we have then the best possible safeguard against its being "thrown out" by frost, an evil which takes place if the surface is too fine. But the gradual crumbling down of the clod already mentioned constitutes an excellent top-dressing to the wheat, keeping it from being thrown out. This shows how the clay land farmer works in harmony with the forces of nature and brings them to his aid.

Quite a contrary picture might be drawn of the farmer who insists in proceeding with his tillages wet or dry, who ploughs his clay land in wet weather, and finds afterwards that it is baked into a sort of sunburnt pottery. In vain he struggles to produce a tilth. It is done, if at all, with immense labour and difficulty, and when so produced it is liable to be undone and destroyed by a sudden heavy fall of rain, which produces a mortar-like condition of the soil.

Another principle which must be steadily kept in view in the management of clay land is the policy of lightening it up—a policy of rendering the soil less stiff and less solid—of disintegration. It is the very opposite to the policy pursued in treating light land, which is summed up in the word consolidation. Light land requires consolidation, heavy land requires the reverse. This policy of lightening up clay land is accomplished in various ways. One is the application of farmyard dung in the green or fresh state; the second is the application of various substances bulky in character, which are meant to mitigate and alter the soil—for example, cinders, ashes, town manure, lime, haulm of various kinds—anything which will tend to lighten up and separate the particles of the soil.

With reference to root cultivation on clay land, these soils are not well adapted for this purpose. Two difficulties beset the clay land farmer with reference to root cultivation. It is not by any means easy in the case of stiff clay soils to secure that fine and moist condition suitable for the germination of the turnip seed. This difficulty may be overcome by careful tilth, and if it can, there is not the least doubt that clay lands will grow fine turnips, swedes, or mangolds. But another difficulty besets us later on, and that is the disposal of the crop—a very serious difficulty indeed upon stiff clay soils. What are we to do with this turnip crop after it is grown? We may have twenty or twenty-five tons or more per acre of material to dispose of. What are we to do with it? If we put sheep upon it we shall find that neither the sheep nor the land are benefited. Sheep will not thrive upon clay land in winter. If we could contrive to feed them off in summer the case would be perfectly different; but sheep will not do on heavy clay land in winter, and the land suffers even more than the sheep. Now we know perfectly well, as an economic fact, that the turnip crop is not often grown at a direct profit. The reason why we grow turnips is that, although they are grown at a loss, yet they prepare the land for the succeeding crops. The following corn crop is grown very cheaply after turnips. It may be barley, which again is followed by clover, grown very cheaply, and that again by wheat, grown at comparatively little expense; so that the turnip crop, although grown at a slight loss, is profitable, taken in connection with the succeeding crops. The turnip crop is a fallowing crop, is a substitute for the bare fallow, and is much cheaper than the bare fallow. We cannot bare-fallow land under £5 per acre, and if we can grow turnips at a loss of £1 per acre, we do neither more nor less than fallow our land for £1 an acre, which is cheap work. All depends on whether we efficiently fallow our land with the aid of turnips. Upon light soils we do so. Upon light soils the turnip crop is in every respect a substitute for the bare fallow, and more than a substitute. It is better in all respects. A turnip crop consumed upon light land at once imports into the soil a mass of organic matter and material derived from the atmosphere, and a mass of available material collected from the sub-soil. It likewise insures the thorough cleaning of the land, and lastly, the texture of the ground is vastly improved by the trampling of the sheep; so that upon light land we have in the turnip crop a great improvement on the old bare fallow, to which we must add a great reduction in cost. Therefore the case is as plain as a pikestaff in favour of turnip cultivation on light land. It is well to note that there must be a good deal of debateable or border land. There must be a good deal of land which is neither heavy nor light, and as to which it may be very difficult to decide whether we should grow roots or not, the question being solved very generally by the character of each season. In some seasons more land will be brought into turnip cultivation. That will be in dry seasons; in wet seasons more land will go out of turnip cultivation and recruit the area under bare fallow.

In the case of stiff clay land, whether we consume the turnips on the land with sheep or haul them off the ground, in either case we damage the land. We grow a turnip crop, and instead of a better series of crops after it, we have a worse. This is a very important point. Upon certain classes of stiff land, if we force turnip husbandry upon it we shall find we grow a worse wheat crop, and a worse succession of crops owing to the poaching and the trampling of the ground in the process of turnip cultivation; so that turnip cultivation is robbed of its greatest attractions, and becomes a difficult and expensive process, followed by inferior results. And when we grow an inferior crop of corn after turnips we may be disposed to think it may be better to barefallow such land and to bring it into wheat. There is another point which touches upon this question, and it is that these clay lands of which I am speaking are essentially wheat lands. They have suffered terribly during the last few years on account of the reduction in value of their main product, but they are wheat lands as long as they remain in arable cultivation. Upon such clay lands it is a point that wheat should be sown early. Wheat ought to have two summers in order to develop thoroughly, or as much of two summers as it can get. The old-fashioned farmers used to sow their wheat early in August, and thought that the young wheat should be peeping out of the ground in time to see the old wheat riding home to the rickyard from the harvest-field. We do not see that now on account of the spread of turnip cultivation. We sacrifice our wheat in some measure to our live stock; but wheat, if it can be planted in August on cold clay lands, such as I am now speaking of, comes up strongly and rapidly, and spreads itself like a green carpet over the soil, and although it may perhaps become winter-proud, yet at the same time we find that early sowing is the way to get a strong-rooted, thoroughly good crop of wheat. But all this is quite inconsistent with turnip cultivation, because if we cultivate turnips it is difficult to get them off the ground before October, and it is equally difficult to get our wheat in before November, which is too late for securing the best results.

Such considerations as these will guide a farmer as to how he should crop his land. In the cultivation of clay land it seems desirable at the present day to extend the rotation beyond that of the ordinary four years' course, which is more adapted for light soils, and these soils, as they yield a good return in grass, may be left down two, three, or four years in a proper mixture of seeds and clover, with very good results.

I pass on to some of the principles which control the cultivation of light land. I used the word consolidation with reference to the cultivation of light land. We desire to produce a firm condition which is natural to clay soils; there is no difficulty in getting a firm condition in clay soils, we rather require to operate so as to produce a looser condition in them, but in the cultivation of light soils we want to produce as far as possible a solid, firm condition. Much of the farmer's attention is concentrated on making his light soil firm. He digs his heel into it, and finds the ground is loose, and requires the roller over it; or he is satisfied if it has a firm feeling under his foot. He must have it fine for the germination of seeds, but he must also have it firm. Especially is this the case in corn growing. If I were going to sow turnips I should like to have the ground a little loose under foot, but if I were going to sow wheat, barley, or oats, I should like to have the ground firm. The advantage of firmness of soil may be illustrated by the benefit which follows the treading of sheep on light soils. It is the very making of these soils. It is the treading of sheep which has enabled the Norfolk farmers to grow corn on unpromising sandy soil. The old Persian proverb that the sheep has a golden hoof is particularly true and applicable upon light soils, but it is not at all true upon heavy soils. We see it also in the system of cultivating wheat on light soils. The land-presser comes into vogue, following the plough; we not only must plough our land for wheat, but we must press the furrow tight, either with a land presser or with the Crosskill roller. It is strange that this roller, which was brought out by Messrs. Crosskill, of Beverley, as a clodcrusher, was never taken up heartily by clay land farmers for crushing clods; but has found its way into the hands of light land farmers, for the purpose of following up the plough, and pressing and "firming" the furrow, after ploughing for wheat.

Plough early for wheat. To obtain a solid, firm furrow is one of the reasons why this is done. Upon light land we wait for rain before we sow our wheat; we will not sow our wheat in a dry seed-bed; we wait for rain for the same reason—that the soil may be firm; and we even go so far as to say in the case of light land, that we like to see wheat smeared in, because we artificially produce in that way somewhat of the same character as we find naturally in a clay soil. Wheat can be put into light land in a very wet state; it may be muddled in, plastered in, and in the spring of the year we shall find it come up strong and well.

I will give you another example. On light land we are in the habit, in my district, of driving our sheep over the young wheat. The reason that the practice has gone out to some extent is the increased value of the sheep; they are too valuable to be used in this way. The older-fashioned farmers of the west country moved their flocks backwards and forwards over the wheat fields with the same object in view—to consolidate and "firm" the ground. Before ploughing up land for wheat the grass should be eaten off as bare as possible, because otherwise it lies too hollow; but if the grass is eaten off bare the furrow packs closely. Again we shall find that farmyard dung on light lands should be applied in the well-rotted state. It should be short for light land, because we do not want the ground to be lightened up with strawy stuff—it lets the drought in.

I have given enough examples to show that light lands are cultivated upon the principle of consolidation. Light land is easily over-cultivated or over-tilled. We may "plough the life out of it"; the strength out of it. Provided we can get light land clean, the less we do to it the better. If light land is clean we may always depend upon its being fine, and if we do not touch it we may rely upon it being moist. Tillage is often carried on at a great sacrifice of "sap" or moisture in the soil; therefore, one ploughing, or even one cultivation, in the case of a piece of clean light land will be quite sufficient, and there is no object in ransacking the ground about for the purpose of dividing it, as is the case with clayey ground.

Now we approach the subject from another point of view, which we have a little anticipated, viz. with regard to the crops which have to be cultivated. Different crops require different systems of cultivation. I will take the root crop first, and go through the rotation. The root crop requires deep cultivation. Mangel-wurzel, carrots, swedes, kohlrabi, rape, all like deep cultivation. They are all deep-rooted plants, and they all enrich the surface of the soil by pumping up material from the deeper layers of the subsoil, as well as from the atmosphere, and when they are fed upon the surface they enrich the top soil, upon which corn crops feed. It is for these crops, therefore, that we chiefly use autumn cultivation. Autumn cultivation is a cultivation in autumn undertaken for the root crop of the succeeding season. Now the root crop is a cleaning crop and a fallowing crop. Therefore, it is at this period of the rotation that cleaning operations take place. The root crop also cannot be grown unless the soil is rich. It must have plenty of available plant food, because it takes such a short period to grow. Turnips may be ready for consumption eight weeks after sowing, and if we take from the 1st of June, when swedes may be sown to advantage, to the 1st of November, when they almost cease to grow, the period is short for accumulating fifteen, twenty, or thirty tons of produce per acre. The plant requires its food in abundance, and ready to hand, so that in root cultivation we must apply plenty of fertilizers. It is a very expensive and complicated cultivation. It is much more expensive, much more complicated, and much more difficult than corn cultivation. With reference to the actual tillages for roots, they must be conducted in such a manner as to provide a deep, moist, rich, and fine seed-bed. Now the cleaning of land in the spring of the year is very apt to dry it, but the cleaning of land in the autumn months does not dry it, because we have the winter rainfall to come. Therefore, if we can clean our land in the autumn, immediately after harvest; if we can plough it up and get the dung in, and get it into furrow previous to winter, we have done a great deal towards securing that cleanness and that tenderness and that moisture which are so highly desirable. Then in the spring of the year we must touch it as lightly as we can, whether it is light land, or whether it is heavy land, because in the case of light land, cultivating it much in the spring loses moisture, and in the case of heavy land, late spring ploughing and much spring cultivation buries the fine surface of the soil, and brings up a great deal of tenacious stuff not fitted for the use of the young plants.

The advantages of autumn cultivation for roots are very considerable indeed. It secures the performance of a work which probably might be very difficult to resume in the month of March or April. We know that clay land is often scarcely fit to stand the pressure of horses and implements at that season of the year. The second advantage is that we break up the homes of entomological pests. Entomology is an important subject for agriculturists to study, but I am struck with the remedies which entomologists propose to us agriculturists. They propose exactly what we should do if there were no entomologists. These remedies generally come in the ordinary course of good farming, and therefore they lose a little of their significance. They tell us we must autumn cultivate. Of course we must; we do that, not for entomological reasons, but for other reasons. We know the habits of our insect enemies; towards the autumn months they penetrate the ground, and make their little cocoons. They line their little cell-walls with beautiful material, and they make up their minds for a pleasant winter. Then the plough comes and turns them all out, exposes them to the ravages of the birds, and to the mercies of the frost; and then, of course, we have a wholesale destruction of our insect enemies. The exposure of the seeds and roots of weeds is another advantage. They also are preyed upon by birds, and are exposed to the rigours of the winter, and there is a general destruction of these vegetable pests.

Next we have the grand reason for autumn cultivation which overrides all the rest, and that is the beneficent effects of the action of the air and the action of frost, and changes of temperature in pulverizing soils. In this connection I may repeat also the conservation of moisture in the spring of the year, when the ground becomes thoroughly soaked with wet in the winter, and that moisture is conserved in the depth of the porous soil, and becomes useful for growing crops in the summer. These are the reasons why autumn cultivation is popular among good farmers.

Now the best method of cleaning a field is a thing which ought to be taught, and which does not appear to have received sufficient attention from teachers of agriculture. I scarcely ever find a candidate select a question asking the way in which he would clean a foul piece of wheat stubble. They prefer to enlarge upon the dormant constituents and the double silicates. They do not like farming so much as chemistry. But as long as I can explain tillages with a certain principle running through the explanation, I think that I am teaching the principles of agriculture. There is a principle in the cleaning of land, and I will try to explain what that principle is. Take the case of a foul wheat stubble, which must be got ready for swede turnips next year. Immediately after harvest we must get to work; I do not care whether the steam-cultivator is used or not, but whether steam or horse implements are employed, first cultivate from north to south, and then cross-cultivate from east to west; in other words, cross-cultivate, so as to hit all the ground. I do not care whether you carry out the system just mentioned or whether you thin plough or pare it with paring-ploughs or with broad shares, about fifteen inches across the blade. The object in all cases is to detach the first two or two-

and-a-half inches of the surface soil which contains the weeds—including the couch. On a previous occasion I mentioned that immediately after harvest the weeds are weak, the couch lies near the surface, and it is weak from the overshadowing of the corn crop. Then follows the drag or heavy harrow, which has no difficulty in penetrating to the bottom of the work. The moved portion is thoroughly knocked about with the heavy drags, once rolled, four times harrowed in different directions to bring the couch out, twice chain-harrowed to separate the couch from the adherent soil, couched, burned, and the ashes spread. That ends the first series of operations. Next plough it, four* times drag it, roll it, four times harrow it, twice chain it, couch it, burn, and spread the ashes. That is the second series of operations. That ought to be sufficient to clean a field. Thirdly, in places where couch and other weeds are most prevalent, we may have to repeat these operations a third time in certain spots, and in such case plough them, give them a good dragging, harrow, chain, couch, burn, and spread the ashes. Two coats of couch very often are brought out, and the third couching will generally be partial only, and restricted to the foulest spots of the field. Then cart the dung on, and give the winter furrow as deeply as is advisable, but not so deeply as to bring up any of those injurious materials or seeds of weeds or other drawbacks, and leave it till the spring. Such is autumn cultivation.

Briefly reviewing the above given cultivations, observe that there is a principle or method running through them. First, the section of the ground containing couch and other weeds is detached from the section below it by the plough, the broad share, or the cultivator. Next this weedy portion is

---

* The numbers of operations here recommended are not intended to be understood as absolute, but are given as fairly approximating to what is usually necessary.—I. W.

thoroughly pulverized by drags, harrows, and rollers. In the third place the weeds are collected, burnt, and the ashes are spread. This constitutes a series of cleaning operations. They must however be repeated, and this is done by a shallow ploughing which is followed by processes contrived to again bring up a crop of couch and other weedy matter to the surface, which again is collected, burnt, and spread. This constitutes a second series of cleaning operations. It is usually now only necessary to subject certain especially foul portions of the field to a third series of operations, and the first part of autumn cultivation—the cleaning of the land—is ended.

The second phase consists in carting on and spreading a coat of dung over the surface; and the final phase consists in covering the dung by a deep winter furrow, which ends the process.

With reference to this statement of tillages, it by no means follows that it is exactly followed, but, at the same time, what has been given is usual, reasonable, and possible. People often treat their land in the manner indicated, but every farmer knows that the amount of cultivation, and the operations which he thinks necessary, must depend in each case on particular circumstances. Still, if as an examiner or as a practical agriculturist examining a young farmer, he were to describe the cleaning of land as above, I should think he had a reasonably good idea of how to clean land.

CHAPTER X.

Classification of Crops—Principles of Cultivation for Root Crops—For Corn Crops—For Grass Crops—For "Fodder" Crops and "Catch" Crops—Syllabus of Crop Cultivation.

WHEN I have the opportunity of entering at length upon the subject of crop cultivation, I naturally take the several crops *seriatim*, and discourse upon all the minute variations of treatment which the exigencies of each crop require. But where space is limited it seems wiser to attempt to give, in one fairly comprehensive view, the principal points affecting the chief groups into which we agriculturists divide our crops. We usually speak of (1) root crops, by which we mean crops cultivated on account of their roots, such as turnips, swedes, and mangold wurzels. (2) Grain crops, which include not only cereals, but also beans and peas, which are grain crops, although they may also be termed pulse crops or black crops, in contradistinction to white or cereal crops. Then we have (3) the cultivation of grass crops, which includes clovers, and all those plants which take a place in mixtures intended for grazing purposes, although they may not strictly be included in the botanical division of grasses. (4) We have lastly fodder crops, which are grown with a view to the leaf and stem being used for feeding purposes, and which usually occupy the ground for only a very short time.

In reviewing the necessary cultivation of these four classes of crops, we take the root crops first, chiefly because the root crops are a substitute for the period of fallow, or the period

THE PRINCIPLES OF ENGLISH AGRICULTURE. 133

during which land is renovated, cleaned, and made fit for a succession of other crops. I have already endeavoured to point out some of the most interesting matters in relation to the cultivation of roots, and the preparation of the land for them. That lead me to the cleaning of the land, because it is most essential that land should be cleaned in anticipation of the root crops. We may depend upon it, if land goes foul into roots it will continue foul until the end of the chapter. There is no other opportunity of cleaning land, and that is readily seen by passing in review the course of what is called the Norfolk rotation.

After roots, the ground is speedily put into barley; upon the young barley you sow your grass-seeds, so that when the barley is severed from the ground the third crop, or clover crop, is in absolute possession of the ground, and it is therefore perfectly impossible to attempt any cleaning operations during that period. Finally, after the clover has laid one, two, or three years, as the case may be, it is ploughed or pressed, well harrowed, and drilled with wheat. It is not often that any cleaning operation is attempted, but only what I have briefly indicated. Therefore, if the ground has been foul at the time when it went into the root crop, depend upon it, it will come out ten times fouler when the wheat crop is taken off. In fact, if ever there is a period in farming in which a little couch and a little foulness of the ground is excusable, it is after the wheat crop has been harvested, at the close of the four or five years' course.

This is the reason that I laid such stress upon the importance of cleaning land for the root crop. After cleaning I recommended the application of farmyard dung in sufficient quantities, to be ploughed in as deeply as we could or as we dared, and it was then to be left to receive the influences of the winter frosts.

The spring work—the spring cultivation—in some cases

includes ploughing, specially upon that large medium class of soils which is fairly retentive of moisture, and is at the same time easily brought into a state of tilth. Upon such land the plough ought to be used in the spring of the year, but repeated spring ploughings and late spring ploughings are not to be recommended.

In the case of very stiff lands and in the case of very light lands avoid ploughing late in the spring of the year. Ploughing is essentially an inversion of the soil, therefore ploughing buries the fine surface, and brings up retentive raw material, whereas cultivation or grubbing, whether done by horse or by steam power—and I prefer the latter—breaks the under soil, but it retains all the fine pulverized material upon the surface. Likewise in the case of light land, late spring ploughing is to be undertaken with caution because it dries the ground to too great an extent. It is impossible to exactly describe spring cultivation; but our object is always the same, namely, to preserve and obtain a fine surface to keep in the moisture, and to secure a thoroughly clean soil, or seed bed. Root cultivation differs from grain cultivation in another respect, namely, that the crops are subjected to a systematic course of after cultivation. As soon as the young turnips are visible hoeing commences, very frequently flat hoeing or hand-hoeing, followed by horse-hoeing between the rows, followed by singling out the surplus plants, setting them out at intervals of ten, twelve, or even thirteen inches from each other. Horse and hand-hoeing are repeated according to the district in which we may be. In the north of England the stereotyped idea is three horse-hoeings and three hand-hoeings—three horse-hoeings between the rows, and three hand-hoeings in the rows. In the south of England, where harvest is earlier, this important operation interferes with the third hoeing of the turnip crop, and twice hoeing is all that the crop usually receives.

ENGLISH AGRICULTURE.

The root crop requires liberal treatment. The greater portion of the farmyard dung produced upon the holding is bestowed upon the turnip or fallow breadth. It is, however, better to put a portion of the dung produced upon the turnip crop, and to reserve a portion for the wheat, placing that latter part on the clover about to be broken up. Especially is this the case upon the lighter class of soils; there I should recommend the farmyard dung to be divided—a portion to be put upon the roots, and a portion reserved for the wheat crop. Then again, with reference to the application of artificial fertilizers, they are now universally applied to root crops. Without artificial fertilizers we in many cases could not obtain a crop at all. Superphosphate of lime is the substance in general use over the whole of the Midlands and Southern England.

When we go north into a cooler climate guano, in a certain measure, takes the place of superphosphate of lime, and in Scotland, as well as in Ireland, guano is in high repute. But over the greater part of England superphosphate is the universal manure for the root crop. In asking the question in examination, as to the proper fertilizer for roots, I am often told that roots contain a great deal of potash, and therefore potash is the proper manure for roots, and my students will often go on to say that wheat contains a great deal of silica, and therefore silica is the proper manure for wheat. Potash is no doubt contained in very considerable quantities in the root crop, but it so happens that farmyard dung contains a great deal of potash; straw abounds in potash, and straw is the material which is almost always restored to the land. It is true that the root crop requires potash, but it gets it in sufficient quantity, and there has never been that demand for potash salts which it was at one time thought there would be. They are used chiefly by amateurs, but the bulk of tenant farmers who have to pay rent neglect potash, at all events with reference to the turnip crop. They use it upon the potato crop, but they adhere to superphosphate in the case of turnips. Superphosphate gives an available quantity of phosphoric acid, and being easily drilled with the water-drill or dry-drill, it is brought into close contact with the young plants, and the effect of the superphosphate upon a young turnip is very extraordinary. One of the chief uses of the superphosphate is to carry the young plants past the stage in which they are only represented by the cotyledon leaves, until they graduate into the rough leaf, when they are comparatively free from the attacks of the turnip fly. In addition the superphosphate exercises an excellent effect in the later stages of growth.

I have endeavoured to show that turnip cultivation is complicated, because it entails the cleaning of the ground, the bringing of the soil into a highly fertile condition, the management of the plant during the whole of its growth. Turnip cultivation is therefore the most difficult, the most complicated and the most expensive of any. A man who is able to manage his turnip breadth will be able to manage the rest of his cultivations, and if he has a sufficient force of horses upon the farm, and of men to manage his turnip breadth, he will have ample force to carry on the rest of the farm. The weakest link of a chain is the measure of its strength; and in the economy of managing a farm the test crop is the root crop.

Next we come to the cultivation of corn. This is a comparatively easy matter. No business pursuit is easier than corn cultivation, and that is why we have such millions of bushels of corn thrown in upon us. It is a cheap cultivation. All we have to do is to plough the land, throw on the seed, and scratch it in. Of course we must do this at the right time of the year, and in a proper manner. When we take wheat, or barley, or oats, after a root crop fed on the land, a very general method of cultivation is as follows—we first plough about four inches deep, then broadcast the seed upon the newly turned up fallow, and put the harrows on and give it a really good harrowing so as to break the compact furrow, and cover the seed thoroughly—that is all. Protect it from the ravages of the birds, and in the spring of the year roll and harrow it, and that is pretty nearly the cultivation of corn after roots. A great deal of corn is taken after grass and clover crops; and the cultivation of either oats, or wheat, or barley after lea is much the same thing. We plough and press, and often sow the seed upon the pressed furrow and harrow it in.

Again, in other cases we plough, press, or heavily roll, harrow repeatedly, and drill. That again is the whole of the cultivation. Corn crops sometimes follow peas or beans, in which case the plan would be to dung the surface, and then proceed as before, ploughing in the dung, and either broadcasting, or else producing a proper seed-bed with the use of the harrow, and drilling in the corn. So that we must regard the cultivation of corn as a simple and inexpensive thing, unless the cultivation of the turnip crop is considered as part of the cultivation for corn, in which case we throw a good deal of expense upon our corn crop; but I prefer to consider the two crops separately.

A very important subject is that of the laying of land down in permanent pasture; but when we speak of the sowing of grass crops in this connection, it is with the view of alternate cropping, or of growing grass for one or two years in a rotation. This is a still simpler and easier cultivation. Clover and grass seeds are sown upon a growing corn crop, usually barley. The fine clover and grass seeds may easily be buried too deeply—a quarter of an inch is quite deep enough. Perhaps one of the best methods of sowing these seeds is to pass the Cambridge roller over the surface, leaving the ground in a series of small corrugations. Let the grass seeds be sown by means of a seed-barrow upon the surface, and lightly harrow them in with a grass seed-harrow made for the purpose, or with a brush-harrow, or with a chain-harrow. This is done in the month of April upon young barley or young corn. The corn grows and ripens, and at harvest time is severed from the ground which the grass continues to occupy.

One of the simplest cultivations I know is that of *trifolium incarnatum,* or crimson clover, a very useful fodder crop, which might be considered in connection with the next group; but I prefer to take it in connection with the cultivation of clovers. This plant is used largely over the southern counties of England, and it gives in the spring of the year a very abundant crop, well suited for sheep. All the cultivation requisite is to give the stubble a regular good dragging or harrowing, say twice in one direction, twice across, and twice diagonally, then wheel in twenty pounds of trifolium, give it another harrowing and a rolling, and that is all. It is simply scratched in. This treatment is much better than ploughing the ground, because trifolium dearly loves the soil to be in a firm state; it likes a fine seed-bed, but it must have the ground firm below.

There is a great similarity between the cultivation of fodder crops and the cultivation of corn crops, many of our fodder crops being in point of fact corn crops. For fodder purposes we grow, especially in the south of England, large breadths of winter barley, winter oats and rye, which are cereals. The ground is ploughed and then harrowed, and the two and a half bushels of seed required are drilled and harrowed in. The other principal fodder crop, namely, vetches, has a very simple cultivation, consisting of ploughing, harrowing, and drilling, and harrowing again after the drill.

Fodder crops are very generally used as catch crops, or in other words, they are crops taken between two main crops, the usual position of the fodder crop being previous to the root crop. The following rotation is practised in many districts in the south of England, especially where sheep-farming is a principal object; first year roots, second year barley, third year clover, fourth year wheat or oats, fifth year winter rye, barley, or oats fed on the land, and followed with roots fed, sixth year barley, seventh year clover, and eighth year wheat. Such a system of farming as this, in which fodder crops precede turnips, must be followed only on the lighter classes of soils, because it precludes the possibility of exposure to the winter's frost, so that stiff land treated in this way would probably plough up tough and leathery, and it would be difficult to bring it into the right condition for roots, whereas light sandy ground breaks up readily after this cultivation, and is speedily got into a fine enough condition for the sowing of the root crops. The system must always be restricted principally to the south of England, for two reasons. First, in the north of England the fodder crop would not be ready early enough. In the north of England turnips must be sown earlier than in the south, so that the two crops will not fit together as they do in the south. There is also a third consideration which ought to be remembered with reference to fodder crops—ground ought to be clean. Notice, there is not much opportunity in this rotation for cleaning operations. The land is for ever under crop, and there is never any lengthened period during which we may set to work cleaning it. My own experience of this system, as practised in a southern county on a proper soil for the purpose, is that it is difficult to keep the ground perfectly clean under it. You never can preserve that perfect cleanness of ground expressed by the statement that "Mr. —— has not a barrowful of couch on the whole of his farm." We cannot do that, but I also believe that it prevents the land from becoming very foul. The ground never becomes very foul, and never is absolutely clean. While the system of catch-cropping is a bar to perfect cleanness, it is likewise a preventative to weeds thriving extravagantly, because the ground is always under tilth, under cultivation. Fodder crops may be grown on stiff land as well as on light land, in the north as well as the south; but where we have these conditions we must make up our minds either for fodder or for roots. We must either grow fodder crop and relinquish root crop, or choose the root crop and relinquish the fodder crop; we cannot do both.

Now I trust I have in some measure, at all events, fulfilled what I hoped to do, and given a bird's-eye or general view of the cultivation of these various kinds of crops; and in the last place, I would confide to the teachers of this subject the syllabus which I employ myself in the teaching of, not only the cultivation of crops, but of all other matters connected with crops. It is extensive, and requires a large amount of knowledge, especially when it is applied to each crop separately.

The syllabus of crop cultivation begins with the *botanical position*—I mean the botanical position of grasses, wheat, clovers, cabbages, turnips, and any other crop. The botanical position of the root crops, for example, is a point which an agricultural student should not be ignorant of.

The *history* of the crop is the second point of the syllabus. It is astonishing how many of our crops have a distinct history; we know when they were introduced and who introduced them. Two hundred years ago we had very few cultivated plants at all. Their introduction forms a very instructive page of history, and must always be an excellent illustration of the value of observation and of science. The history of cultivated plants as well as the history of the improvement of animals are grand chapters in the national progress, and it is well to impress them upon young people, as it may lead them to endeavour to add to the stock of knowledge already handed down.

Cultivated varieties.—This is a very extensive field for inquiry and for instruction—cultivated varieties, with all their peculiar adaptation to different soils and different climates. They throw light on a number of biological questions connected with evolution, and with the relationships between cultivated plants and wild forms. Besides, it is of great importance to know that there are many kinds of wheat, and many kinds of turnips, and that some are suitable for highly-farmed land, while some are suitable for badly-farmed or exposed weak lands. The variation is endless, and is constantly increasing, and I would impress this section of the syllabus upon teachers as of importance, and as highly suitable for purposes of instruction.

The *place in rotation* of any crop you choose to name, say turnips, clover, or beans, is always a subject of interest, and ought to be systematically taught.

Soils suitable.—Students should understand that crops should be suited to the soil. Take the *trifolium incarnatum*, for example. It will not grow at all upon white chalky soil, whereas it will give a very vigorous and luxuriant produce upon the proper sort of soil, namely, a loamy gravel. It would be endless to enter upon this subject in detail, but certainly the soil suitable for the different crops opens out questions of great importance.

Preparation of the ground.—This point we have already considered at some little length, but when the special preparation for each crop is taken separately, we have before us a large subject well worthy of the attention both of teachers and students.

Period for sowing and methods of sowing.—Both of these topics are most important, and both may be taught to classes. The successful growth of crops greatly depends upon the period at which they are planted, and the season for sowing again depends upon soil, climate, and the variety it is intended to cultivate. Thus we have autumn and spring wheats, late and early potatoes, late and early trifolium, and it would probably be difficult to name any crop which does not boast of varieties suitable for sowing at different seasons of the year. So also with regard to methods of sowing. Much may be said upon the relative merits of drilling, broadcasting, dibbling, ploughing in wide and narrow intervals, and the proper depth at which seed should be deposited.

Quantity of seed is a topic on which volumes have been written, quite an acrimonious discussion having been waged between "thick" seeders and "thin" seeders. Some have gone so far as to consider one or two quarts of wheat enough to seed an acre, while others have held that three or four bushels is not too much. The quantity of seed is a very important matter; it varies with circumstances. Upon the subject of the quantity of seed alone a capital pamphlet full of valuable matter might be written, so that here again we have food for sound agricultural teaching.

The *fertilizers* to be employed is another subject sufficiently important to merit a separate chapter to its consideration.

The *after-cultivation* of the crops is an important matter—whether as regards root crops or corn. Operations such as horse and hand-hoeing, rolling and harrowing, top-dressing, and guarding against insect attacks are best considered under this heading.

Securing or harvesting.—The subject of ripening or maturation, and the best stage at which to cut, or to store, or feed off each crop, is considered in this section of the syllabus, also the best methods of harvesting, which leads the teacher to consider the respective merits of self-binders, reapers, scythes, and sickles for cutting the corn, the management of corn in the field, and the proper time for carting it home. We may also be led to consider the question of hay-making, and the principles which ought to guide the farmer in this important task. The securing of root crops is another topic involving many questions of scientific as well as of directly agricultural interest.

Preparation for market or for home consumption.—This may be considered as too completely a matter of practice to form a part of instruction in the principles of agriculture. It, however, gives an opportunity for explaining some points of importance, such as the various instruments required for thrashing, winnowing, screening, and grinding corn, and the use of chaff-cutters, root-pulpers, and cooking apparatus, in the preparation of home-grown produce for stock feeding.

Insect attacks.—A description of the many attacks to which our crops are subject is a topic which is both important and interesting. No sort of instruction could well be suggested combining so well practical utility with the educational advantage of precision of description, and the encouragement of observation. The insects are numerous, and their appearance, life history, habits, and the methods for destroying them all offer excellent material for instruction.

Diseases.—No more interesting or varied subject than that of the diseases of our cultivated plants could well be selected for study. Mr. Worthington G. Smith's book, entitled *Diseases of Field and Garden Crops* (Macmillan & Co.), is an excellent text-book on the subject.

Composition and Properties.—This takes us directly into the domains of physiology and chemistry. The relative value of clover and meadow hay, and of wheat, barley, and oat straw; the relative value of turnips, swedes, yellow turnips, and mangold and of barley-meal, and bean-meal as foods for different kinds of live stock at different seasons of the year, opens up a varied subject of great interest. The variations in composition and in nutrient value at different stages of growth is another section of the same subject.

Cost of production and realization.—Although somewhat outside the scientific aspect of agriculture as usually described, there are reasons for including cost of production and return in a system of agricultural teaching. It belongs to the domain of economy and statistics, and it is of great importance that the cost of growing crops and the value of the produce should be placed before pupils. The interdependence of our crops in a rotation is well brought out in this section. Thus root and fodder crops are often grown at a direct loss, but an indirect profit, inasmuch as they serve as a preparation for saleable corn crops. The difference between market value and consuming or feeding value is also a point which should be explained in this connection.

The seventeen sections above enumerated might be further increased. If each crop, however, is taken separately, and considered in each aspect, a large-mass of information will be brought before the student, of both practical and theoretical interest, and the teacher will remain properly within his province as an expounder of agriculture.

This is all we can spare time for with reference to crop cultivation. If I were to follow my inclination and plunge into details I should miss my present mark. But when you as teachers come to deal with classes, information of the sort above indicated is the sort of information which will be of the greatest value to them. It is a matter of regret that such instruction has not long since been made compulsory in rural districts. Here is a subject at once educational and technical. In village schools how much better would it be than teaching out of the way facts as to the length of African rivers, the heights of mountains, and the habits of barbarians, if the children were taught to know the principal varieties of cultivated crops, with their enemies and their friends, the best fertilizers, the proper cultivation of the ground, the best methods of sowing seeds, and many other matters relating not only to plants, but to animals. Such teaching would interest the children, and it would do much towards producing a greatly improved class of labourers in the country. The labourers are the great barrier to agricultural improvement at the present day. This I know from my own experience. It is a very difficult thing to introduce any novelty. A master may be perfectly convinced of the value of an improvement or of a new implement, but the probability is that as soon as he brings it home his labourers declare against it, and it is henceforth condemned and thrown aside. You may order out that implement and use it, and of course your order is obeyed, but never again is it voluntarily used until the order is repeated. The agricultural labourer is very conservative with reference to his business, whatever he may be in his politics; and what he of all things hates is any proposed improvements in the operations which he is daily performing. This is why I strongly hold that it would be most beneficial that these subjects should be intelligently taught to children in all country schools.

CHAPTER XI.

Fertilizers—Relation between Mechanical and Chemical Methods of Land Improvement—Liebig's Views—General Manures—Farmyard Manure—M. Villes' Views—Rothamsted Results—Why Farmyard Manure is Esteemed—Its Cheapness—Artificial Manures—Defective Teaching as to Manuring—Effects Dependent upon Condition of Soil—And of Climate.

OUR next topic is that of fertilizers. When speaking of the methods of improving land I pointed out that we might approach the improvement of the land from the mechanical or textural side, that is to say, by developing the resources inherent in the soil, and by making up its deficiencies in composition by the addition of extraneous matter—that is, manuring. Now there is a link connecting the two methods, as is often the case, namely, that of chalking, marling, or claying soils, as means for definitely adding to their wealth in plant food. So it is with liming, which is an important means of increasing the fertility of soils. It is difficult to say whether the benefit from liming is more due to the addition of lime or to a palpable improvement which lime effects in the texture of the land to which it is applied.

So also in the case of farmyard manure, which is the typical manure of most farmers. We must recognize the fact that farmyard manure acts partly as a direct fertilizer or increaser of plant food, but also by improving the mechanical condition of the ground. I may remind you of the important effect of farmyard dung in mitigating and altering the retentive character of clay soils, and how important it is on such soils that the dung should be applied in a strawy or fresh condition, so that the full advantages of its mechanical effects may be obtained. So important in fact are the mechanical effects of farmyard dung that Jethro Tull, who has been called the father of modern husbandry, and who flourished in the last century, absolutely refused to believe that farmyard dung exerted any but a mechanical effect, and in his homely phraseology he said farmyard dung fed the plant in the same sense that his knife and fork fed him. It divides the food and prepares the food for the inception of the materials by the roots, and therefore he considered that while the effects of farmyard manure were undoubted, yet at the same time they resembled tillage effects; and he was led from that to the conclusion that all that farmyard dung did, could be done by means of repeated and constant tillage, and, what is more, he did it. That was his theory, and it is a good example of how it sometimes happens that men with defective theories will arrive at a sound practical conclusion. Jethro Tull abandoned the use of farmyard dung, tilled his ground, and produced wheat year after year upon the same land without the use of dung, but merely by the aid of tillage implements.

It is rather curious, after the lapse of many years, and after much light had been thrown upon the science of agriculture, to find in one of his latest works that great pioneer, Baron Liebig, taking up a very similar view to that of Jethro Tull, not only with regard to farmyard dung, but to nitrate of soda, sulphate of ammonia, and common salt. Liebig, as you know, espoused the mineral theory, and according to his teaching the ammonia and nitrogen of the plant were derived from the atmosphere, and therefore all that was required was to add the mineral plant food to the soil, and that being done the plant was able to reach forth its leaves, and take its nitrogen from the atmosphere. When it was clearly demonstrated that the effect of nitrogenous manures were most potent, and exceeded very much the action of mineral manures, then Baron Liebig said that the action of nitrate of soda and sulphate of ammonia was not a true fertilizing action, but that they acted as digesters on the mineral matters in the soil, and caused certain mineral matters to pass into a state of solution, in which the plant might take them up readily. But Baron Liebig has passed away, and while we honour his memory, we have at the same time modified a great deal of what he taught us. We now believe that plants take their nitrogen in a very great degree from the soil, and that nitrogenous manures are among the most potent and most active fertilizers that we have. It is a difficult matter to give in a short space the cardinal points on such a large subject as that of manures or fertilizers. It is very much easier to take them in detail and discourse at length upon the merits or demerits of each of the various fertilizers presented to us in the market. But that is not my aim at present. I want rather to draw attention to some of those general principles which underlie the application of manures. In doing so it is necessary to make a decided distinction between two great classes of fertilizers, namely, general manures and special manures.

A general manure is not a general purpose manure, and a special manure is not a manure used for a special purpose; but a general manure is a manure which contains all the constituents which the plant requires, and all the constituents which are removed from a soil by a growing crop. A special manure, on the other hand, is a substance which contains one or more constituents the plant requires, but certainly falls far short of the complex constitution of what is called a general manure. General manures are almost always either of animal or of vegetable origin. As an example you cannot do better than take farmyard dung, derived as it is from straw, from haulm of all kinds, and likewise from the remains of food consumed by animals, and it contains all the effete matter which is eliminated by the kidneys, and therefore all the effete tissues of the body. All the effete matters of the blood and bone find their way into farmyard dung. It also receives all the waste and surplus matter which passes through the alimentary canal, the fæces or dung, and urine of animals, which of course clearly point it out to be highly complicated, and to contain all those substances which go to the nutrition of the body and the formation of blood. It is produced not only from straw, but from grain, and all the numerous cakes—such as linseed-cakes, cotton-cakes, and rape-cakes. It may be regarded very properly as the remnants of grain, or at all events of seed. Farmyard manure, therefore, is a substance which contains everything which is likely to be removed from the soil, or which is likely to be required by the plant. There are not many substances which can compare with farmyard manure in this particular, but among them I would name the following : Town sewage has much to recommend it, and yet it has much which detracts from its value, especially the vast amount of water which is associated with it; but inasmuch as town sewage is the excreted matter of a human population feeding upon highly complex foods, so far does town sewage resemble in general composition farmyard dung.

To name a few more examples. Take guano, or the droppings of sea-birds. Peruvian guano has the advantage of being accumulated in a dry climate, where rain seldom falls, and consequently we have all the soluble materials retained which belong to the fæcal matter deposited by the birds. In many respects guano may be likened to farmyard manure. It is short of one element, potash ; but with that exception guano may be considered as belonging to the type of general manures. Refuse cakes, which may be applied directly to the land, instead of being passed through the digestive system of farm stock, may be considered as general in character. The refuse of slaughter-houses, giving the basis of that well-known manure, Odam's blood manure, ought to have most of the characters of a general manure. The refuse from glue-works, refuse from fisheries, refuse from tanneries, from woollen factories, the refuse which is brought over from South America, as the small bones with flesh adhering to them of slaughtered animals—and the list no doubt could be prolonged. Any substance which is the refuse of either animal or vegetable life will be general in its composition. Compost heaps, or collections of rubbish of all kinds, including occasionally the bodies of animals, mixed with earth, give a rich compost, which owes its value to the fact that it is a substance containing all the requisite materials. Another instance is that of seaweed, which during its life collects its nitrogenous matter, its alkalies, its phosphates, its chlorine, its magnesia, and other substances from the surrounding water. Finally, it is torn up in stormy weather and stranded upon the coast, and forms a very excellent general manure. River weeds also may be employed as a fertilizing agent.

In treating of these general manures, I shall take farmyard dung as the type, as there is no substance that has been introduced to the notice of agriculturists which has superseded it in their estimation. Farmyard dung in this country at least holds its own, in spite of M. Ville, who has written severely against its use, and who considers that it is the most expensive of fertilizers; in spite also of certain results which have been obtained in this country, and which are certainly very extraordinary. Any inquirer may, I feel sure, enjoy the opportunity of verifying what I am now about to state at Rothamsted or Woburn, where a valuable set of experiments are being carried out on the Duke of Bedford's farms. At Woburn, during the last eleven years, farmyard manure does not contrast in its effects very favourably with certain combinations of artificial manures. It is a startling sight to see a definitely heavier crop of corn growing at Rothamsted or at Woburn after manuring with artificially produced salts—it may be of ammonia salts, or ammonia salts plus certain other substances, but nevertheless of artificial salts, than with repeated and constantly repeated dressings of farmyard manure. With reference to these results, I cannot say that they have shaken my faith in farmyard manure, because there is a very great difference between consecutive corn growing, coupled with consecutive dunging of the ground, and the ordinary system of farming by rotation of crops. When wheat or barley is taken year after year off the same land, there is necessarily a long period between harvest and the spring of the year, when the crop is again sown, during which the ground is unoccupied, and during the whole of that period nitrification is busy, and the materials which compose the farmyard dung are subjected to decay, and to washing through the soil, whereas in the case of ordinary farms, where the stubble is at once ploughed up and planted with rye, vetches, trifolium, or other fodder crops, or it may be where you have a carpet of clover or rye grass succeeding barley, there you have a different combination of circumstances—nitrification proceeding under a network of roots removing the nitrogenous matter as fast as formed, and accumulating it in the surface layers of the soil for the further use of the plants. That is a different state of things altogether to that in which we have a repeated crop cultivated upon the same soil. Also it is scarcely just to farmyard manure that it should be exposed to the action of the same crop year after year. A variety of crops introduces a new element, and so far as practical observation goes in ordinary farming, there is perhaps no substance which gives such satisfactory results to us as good farmyard dung. I think it worth while therefore to enlarge a little upon the reasons why it is that farmyard dung should hold such a strong position in the estimation of the British farmer.

The first reason no doubt is what has been already advanced—the general composition of dung. A great many science students stop here. When they are asked why farmyard manure is a more potent and more valuable manure than many artificial fertilizers, they say it is because of its general composition. But there are a good many other reasons beside, one of which is no doubt its effect upon the mechanical condition of the soil, a subject which we have already had before us, and which it is therefore not necessary to further enlarge upon. Then, in the third place, there is the reaction of the carbonic acid gas which is evolved from farmyard dung, upon the mineral matter in the soil. I do not doubt in the least that it digests the soil.

I do not doubt that Jethro Tull was perfectly right when he said farmyard manure prepared plant food. No doubt it does; it is the source of carbonic acid gas, and we know that that gas in watery solution reacts on the mineral matter in the soil with great effect.

Now take another reason. Farmyard dung is rich in nitrogen; that alone places it on a superior basis to most artificial manures. It is rich in nitrogen in a state of organic combination, from which it is liberated slowly by the process of decay, that liberation of nitrogen being known as nitrification. Performed under favourable temperatures, with access of air, and no doubt also assisted by the agency of certain bacteria which work in the soil and produce the peculiar fermentation necessary. This nitrification of farmyard manure in the soil is arrested at freezing-point. It proceeds very slowly at low temperatures, and with accelerated speed at higher temperatures. Especially does it take place freely during the summer months, when vegetation is most luxuriant.

Hence farmyard manure subjected to gradual decay yields up its materials, especially nitrogen oxydized into nitrates, at that period of the year when they are wanted. It is worth notice that the same forces which liberate nitrogen must also liberate the mineral and other constituents of farmyard dung, gradually and as required. I remember being taught as a student, that all substances are in the most active state when they are in a *nascent* condition, or newly liberated from combination. I am not sure if this doctrine is still taught; but if such is the case, it is one reason why farmyard manure exercises such a superior and long-continued effect on growing vegetation compared with the effect of artificial manures.

The durability of farmyard manure may be looked upon as one of the reasons why it retains such a great popularity. That property is well known. It, of all substances, gives what we may term cumulative fertility to the farm. Most fertilizers are one-crop manures, or one-year manures—their effects disappear after the first year, and they have very little residuary effect—surprisingly little; but the residuary effect in farmyard dung is one of its greatest points; it seems to tell at least for twenty years. I remember at Rothamsted seeing unmanured plots of wheat, part of which had been manured twenty years ago with dung, still showing a superiority over the continuously unmanured plots which had not been manured twenty years ago. This quality is sure to give farmyard dung a high standard of value for the farmer, especially as farmers often occupy their farms for long periods of time, and sometimes from generation to generation.

There is another consideration with reference to farmyard dung, which ought not to be omitted, although it is a commercial reason, and scarcely belongs to its function as a fertilizer.

Farmyard manure ought to be produced at a very low sum, in spite of what M. Ville says; in fact, it ought to be produced for nothing. It has been the stereotyped opinion amongst farmers all my lifetime that beef and mutton pay through the dung which they produce. The Norfolk farmers are so vulgar as to speak of sheep and cattle as "muck machines"; and to put the expression in more agreeable form, it has always been considered that if cattle leave their dung behind them, that is a sufficient profit. The animals themselves are considered not to be or to be barely profitable, but if they can pay their way, and leave their dung behind, that is considered as a *quid pro quo*. In other words, if you can get your farmyard dung for nothing, that is a sufficiently satisfactory result from the fattening animals. Years ago Sir John Lawes and Dr. Gilbert pointed out that beef and mutton and pork are all produced at a slight loss, and that the *raison d'être* for carrying on the feeding of cattle and sheep lies in the fertilizing matter which they leave behind them.

I have examined very accurate accounts kept on large estates in South Hungary, where culture is carried on in a much more systematic manner than in England, and there I saw elaborate calculations, showing that farmyard manure was not quite produced for nothing, but nearly so. It was produced at a very small—a nominal cost. If however we carry on our agriculture in a right manner, farmyard manure must be looked upon as a bye-product. If, for example, we breed high-class animals worth more than beef and mutton prices, if we enter the arena as successful breeders of sheep or cattle, and sell the animals for a much higher average than ordinary beef and mutton, then the breeding of these cattle becomes a profitable business, and we may look upon farmyard manure as a found material, a bye-product, and as not costing us anything. The cheapness of farmyard dung, when farming is carried on successfully, is no doubt one of its greatest attractions to the farmer.

Farmyard dung cannot be applied wrong; it is good for every crop, whereas there is room to make a great many mistakes in the purchase and application of artificial manures.

manures. Farmyard manure is also a thoroughly genuine substance. We know what it is; and when it is made under good conditions, it is difficult to imagine anything more fitted for keeping up and adding to the fertility of the soil.

Pass now to the consideration of some of these special manures. There is a large number, but we can classify them into two principal groups. One is the phosphatic group, and the other the nitrogenous group. Plants require, and take from the soil a considerable number of substances, but most of these substances exist in the soil in inexhaustible quantities, and all efforts to persuade farmers to use gypsum, magnesia, and potash salts have failed up to the present time.

It is in very few instances that the substances last named are applied to the land with any marked effect whatever. Phosphoric acid and nitrogen are two substances which are not more important to the plant than any others, but as they occur in the soil in too small quantities for the requirements of the plant, they are always welcome. Phosphatic and nitrogenous manures are the two principal restoratives which commend themselves to the agriculturist, and phosphatic manures appear to exert a most marked and positive effect upon all the cruciferous crops.

Cabbage, turnip, swedes, rape, mustard, kohl or kale, all belong to one natural order, the Cruciferæ, and are particularly affected by the use of phosphates. On the other hand, our graminaceous crops, such as wheat, barley, oats, rye, and all the grasses, are at once affected in a marked manner by the application of nitrogenous manures; so that as a general principle we are led to prescribe phosphatic manures for root and fodder crops, and nitrogenous manures for cereals and grass crops. In this manner, during the rotation of crops, the ground is enriched by both these important fertilizing elements.

I have noticed a defect with reference to the prescriptions given me as an examiner for the fertilizing of various crops. I asked a question during the last examination in which the student was required to prescribe fertilizing substances for turnips, for wheat, and for potatoes. In the majority of cases the question was answered as follows:—"Wheat requires phosphates and silica," therefore give it phosphates and silica (which was a very reasonable deduction); "turnips require a great deal of potash, therefore give them potash; potatoes require potash, therefore give them potash." But it is worthy of notice that because a plant requires certain substances it by no means follows that those are the right substances to give it in the form of fertilizers. Wheat, it is true, requires a large amount of phosphates, but the long period of time which it occupies during its growth, extending perhaps from September of one year to August of the succeeding year, during most of which time the roots of wheat are active in their search for plant food, gives it every opportunity for getting phosphates. It has quite sufficient time and quite sufficient root surface for obtaining all the phosphates it requires out of the soil, and that being the case it does not thank you for phosphates. But it is not so with nitrogen. It wants the nitrogen in larger quantities, and it appears to require its nitrogen more ready to its hand, probably because it is shallow-rooted; but whatever the theoretical reason may be, the application of nitrogen to a corn crop causes a very rapid growth, and it enables the plant to well push its roots through the ground, and to absorb all the phosphates and other mineral matter which it requires.

Another point of some importance with reference to the use of fertilizers is this, that the effect of a fertilizer depends more upon the condition of the soil than upon any other factor. This renders the subject of fertilizers difficult; it renders any recommendations which we might make possibly misleading. It tends to render the reports of agricultural experiments useless, because so much depends upon the nature of the soil. Upon some classes of soils you will find great effects from gypsum, but if you were led to purchase gypsum on a large scale because you read of experiments favourable to its use, you might be easily disappointed. It is the composition of the soil which affects the use of fertilizers more than any other thing, and we ought to use our chemical knowledge in elucidation of this point. If a soil is deficient in lime, it wants lime: and if the soil is found to be deficient in magnesia, or in potash, of course we have an index pointing the direction in which fertilizing may be of benefit. If a soil is open in texture, it is extraordinary how small a proportion of a substance will suffice for the requirements of a crop; but, nevertheless, if a soil is eminently and pointedly deficient in a substance, that is a very good reason for trying that substance.

But there is another view of the case which is still more important, and that is the great influence which the condition of the land has upon the effect of fertilizers. The condition of the land appears to be almost everything, and the *dictum* is that the higher the condition of the land the less effect will a fertilizer have upon it. Poverty-stricken land will respond freely to the use of fertilizers, but land which has been long well farmed and well dunged scarcely responds at all, and very frequently gives absolutely a *minus* result, which is very discouraging indeed. We who have tried experiments for a long series of years, know perfectly well that it is not at all an unusual thing to find three or four hundredweights of superphosphate give less result than is obtained upon an unmanured plot. That does not astonish us at all. It seems to have astonished Sir Thomas Dyke Acland very much, but it does not astonish old experimenters at all when they get a *minus* result after the application of thoroughly credited, well-established fertilizers. It is not anything against a fertilizer at all, but in conducting agricultural experiments we must always be guided by contrast or by comparison, and if our land is in such condition that our unmanured plot gives a high result, of course our manured plots give a less showy or a less marked result than if we have a miserable failure on our unmanured plot. On bad land, or land that is out of condition, our unmanured plots yield very little, the ground being in a state of depletion, and where we have applied the fertilizer, there we see a great effect. M. Ville, when making experiments on the values of artificial fertilizers, did well to select the very poorest soils that he could possibly get hold of, and soil that gives a minimum result on an unmanured plot is sure to respond well to fertilizers and to give a good comparative result. This is an important fact to be remembered by agriculturists, simply because the farmer who has his land in high condition would do well to pause before he purchases fertilizers. They pay best where we enter upon a thoroughly worn out farm, but if we enter on land in high condition it is a question whether we will not throw our money away upon bought manures.

There is another aspect of this question which I wish to open up, and that is the climate. A moist, cool climate is the climate for exhibiting the effect of these artificial substances, and as we travel southwards and approach those regions where a hot and droughty summer may be expected, artificial manure ceases to be of constant value. Thus, in the south of Europe, I understand that it is only once in several years that artificial manure produces an effect, but dung continues to produce an effect. In the hotter portions of Europe we shall find that dung, especially dung produced with artificial food, such as oil-cakes, holds its place, and can be relied upon, while lime, nitrates, nitrate of soda, superphosphates, and other artificials, have no effect at all, or very little effect. That is certainly another point in favour of farmyard dung. If we contrast the climate of Scotland with that of the south of England, we will find that we can apply much larger quantities of artificial manure with good effect in the north than we can in the south. We can apply half-ton to an acre of well compounded substances, as, for example, dissolved bones, guano, and superphosphate. On asking a Northumberland farmer of great experience why he used very heavy artificial dressings for his turnip crops, his reply was that he could grow them in that way at the least cost per ton. But in the south of England two or three hundredweight is quite sufficient I do not know that we can do much better than apply from two to three hundredweight of superphosphate; it is sufficient for the requirements of the climate. Now in Scotland the climate is capable of producing fifty tons per acre of turnips, but in the south of England from ten to twelve or twenty tons is more like what we can produce in favourable circumstances. Therefore the capabilities of the crop in a suitable climate enable it to take up a vast amount of fertilizing matter. The turnip crop has not got the constitution to do it in the south of England. The case is this: in Scotland, where turnips can be grown to a weight of fifty tons to the acre, they will take up any amount of materials we choose to give them; but in the south, where we only look for ten to twelve or twenty tons, it is no use putting on heavy dressings. But in the south of England we may manure mangold as much as we like, because there we have a crop suitable for the climate, and capable of assimilating whatever is given it.

CHAPTER XII.

Action and Re-action of Farmyard Dung throughout a Rotation—Disadvantages from Use of General Manures—Objections to Use of Special Manures—Permanent Effect of One Crop Manures—Principles in applying Special Manures—Allegorical Expressions to be avoided—Effects of Phosphatic and other Mineral Manures upon the Principal Farm Crops—Similar Effects of Nitrogenous Manures—Rothamsted Results—Effects of Fertilizer on Grass Land—Basic Cinder—Principles in conducting Field Experiments.

REVERTING to the division of fertilizing substances into general and special, great advantages may be derived from each class of fertilizers—the great point in favour of general manures is that they are calculated not only to keep up, but increase, the entire amount of available plant food in the soil—they are suitable for all kinds of crops, they are genuine, and less liable to adulteration than many purchased or artificial manures. They are reliable when used by unskilled and unscientific people. Bear in mind that general manures were typified by farmyard dung, and the list of general manure given in the last chapter was mostly composed of refuse animal or vegetable matters. I adhere to my belief in the importance of farmyard manure, and, although experiments at Rothamsted and at Woburn have indicated that artificial fertilizers may be employed with equal effect, and in some cases superior effect, yet at the same time we must remember that those experiments are made in a strictly defined manner, and for a particular purpose. They are carried out upon the same crop occupying the same ground year after year. Therefore we have a different condition of things to what we meet with in ordinary farming. In continuous corn growing there are precisely the same requirements on the part of the plant; there is precisely the same character of root distribution; there is precisely the same period of active vegetation, and in this respect such experiments differ very materially from agriculture as ordinarily carried out.

Let me remind you of the immense effect produced by farmyard manure when applied to the root crop, whether mangel-wurzel, turnips or swedes, and let me also point out that an examination of the results obtained at Rothamsted will show weakness in the matter of the application of various fertilizers to the leguminosæ, including the clovers. It is exceedingly difficult to produce a definite result upon the leguminosæ by the use of any phosphatic, potassic, or nitrogenous application, and yet I appeal to any good farmer in England to bear me out in saying that a thoroughly good mucking will produce a most marked effect upon the clover crop. I have seen it many times, and although it may be difficult to produce figures to prove it, yet I have not the least doubt that if you want a really heavy cut of red clover, or of any other clover, the way to do it is to have plenty of dung below it.

Let us briefly consider the structure of a rotation of crops. Roots followed by barley, barley followed by clover, clover followed by wheat. The first of these crops is greatly and especially benefited by applications of dung; the good crop of turnips or swedes so obtained when fed upon the land is sure to be followed by an excellent crop of wheat, barley, or oats, as the case may be. The fertility left in the ground, subsequently secures a good crop of clover. The clover roots descend deep into the soil, the clover stems rise above the soil. You cut your clover for a hay crop, and after feeding off the second growth plough it up for wheat, and you find your ground well stocked with clover roots. Now, whether recruited from the air ocean, above ground, or from the subsoil, or from both it matters not at present. You have plenty of nitrogen stored up in the first few inches of the soil; the consequence is a heavy wheat crop. That is the way in which farmyard dung acts and re-acts, echoes and re-echoes, during the whole of the rotation.

There are some disadvantages from the use of general manures, the chief being that they store up in the soil certain materials to an unnecessary degree. You add everything, and perhaps do not require everything; there is therefore an accumulation, it may be, of lime or potash, in the soil beyond the requirements of the plants.

Special manures may likewise be objected to, and they also have their strong points. A special manure may contain only one important constituent of plant food, as for example, nitrate of soda. The soda is comparatively useless, because there is no substance so universally distributed in nature, and there is always enough of it in a soil for crop requirements. Nitrate of soda, therefore, boasts one active constituent only, namely, nitrogen in the form of nitric acid. Superphosphate may be considered as an example of a special manure with more than one constituent, for it contains phosphoric acid, lime, a very appreciable amount of sulphur, besides other constituents. But it is wanting in nitrogen and in alkalis, and therefore it is a special manure, but a special manure containing more than one important plant constituent. Lime may be regarded as a special manure, although like all natural products containing impurities some of which are of manureal value. Common salt is a special manure, the chief constituent being chlorine. Sulphate of potash, sulphate of magnesia, sulphate of ammonia, silicate of soda, chloride of ammonium, gypsum, are all special manures. The disadvantage of a special manure is that it tends towards exhaustion, although that result may be very remote. If we continually apply nitrate of soda to the soil, and increase our crops thereby, those crops must search for and appropriate the mineral constituents of the soil, and hasten both its immediate and ultimate exhaustion. This is an objection which is sometimes urged by persons not conversant with the practice of agriculture. But in the ordinary course of farming, what with the use of farmyard dung, the importation of food for live stock in the form of cakes, purchased corn, maize, and hay; and what with the rotation of manures employed, which is as much a fact as the rotation of crops, we think in the ordinary course of good farming there is not any fear of special manures being used in such a manner as to rob the ground to any injurious extent.

There is another view of the case which is worthy of your attention. Special manures, although tending to exhaust the soil, and also somewhat evanescent in their effects, may be so employed as to produce a permanent increase in the fertility of a farm. By the application of nitrate of soda and superphosphate, both of which are special manures, we greatly increase our yield of straw, and of our hay and root crops. When these crops are consumed by live stock upon the farm, they add largely to the amount of dung which is produced. Therefore, in the course of husbandry the comparatively temporary effect of these manures is converted into the more permanent benefits, whatever they may be, which follow the production of large quantities of farmyard dung.

The value of special manures is most apparent in the following circumstances. First of all, where you have a soil which either from analysis or other indications you believe to be deficient in a certain constituent—for example, lime. In such case an application of lime would be of benefit. By the aid of chemical analysis you might detect the want of sulphuric acid in the soil, which would lead you to employ a cheap source of sulphuric acid, such as gypsum; or you might find a deficiency of magnesia, which would lead you to use sulphate of magnesia; or you might find a lack in organic matter, or in nitrogen, or in some other material which you could remedy by liberal application of special manures containing the missing substances.

In the next place, special manures are needed for certain crops. Cereals long for nitrogenous manure, root crops long for phosphatic manure, potato crops cannot be grown to perfection without abundance of potassic substances in the soil. These substances may occur in the soil in sufficient quantities for the requirements of other crops, but not for the crops mentioned, and as one more example, we may take mangel wurzel. The parent form of mangel wurzel is a maritime plant, the *beta maritima*, which grows wild near the coast in situations where chlorine in the form of chloride of sodium is abundant. It is well known that while even in inland districts some twenty pounds of chloride of sodium per acre is yearly brought down in the rain-fall, nearer the coast where sea-frets are common, a very much larger quantity is yearly poured down over an acre. The mangel wurzel being a cultivated form of *beta maritima*, appears from long usage to require a large quantity of common salt, and the application of this substance increases the yield by many tons per acre, especially upon soils of light loamy character. These cases seem to show that special manures are of use in a manner quite distinct from soil requirements. We have, in fact, now before us two explanations of the use of these special substances to which I shall add a third.

We hear nitrate of soda and certain special manures spoken of as "whips" and "stimulants." I do not much like the use of these expressions, which savour of allegory, and are therefore probably misleading. The real fact is, that certain special manures develop, or bring out the powers of the ground. By allowing a crop plenty of nitrogen we enable it to seize upon a larger proportion of mineral matters. By adding to a soil any constituent in which it is deficient plants are enabled to develop thoroughly, and to take up the remaining materials which are present in abundance in the soil. This I think must be evident from the law of *minimum* which was first enunciated by Liebig. If a plant cannot get all the necessary constituents it remains puny in growth, but if we supply the missing link the vigour of the plant is restored. There is no doubt, that when a soil is thoroughly stocked with available mineral matter, by supplying nitrogenous matter the plant thrives amain; it can then easily obtain every constituent which it requires from the ground, and thus the use of the special manure consists in enabling the plant to realize the latent wealth of the soil. This is the way I should prefer to explain the fact before us, rather than by using allegorical phrases.

I shall now give a few illustrations showing the preponderating importance of phosphatic and nitrogenous fertilizers. It would scarcely do to dismiss the whole subject of fertilizers by merely stating that it is only necessary to put plenty of phosphates and plenty of nitrogen in the soil. That would be reducing an important study to a very trite conclusion. Nevertheless, the farmer who keeps steadily in view that he must give his corn crops nitrates and his turnip crop phosphates will not get very far wrong. But in examining the composition of plants, and in glancing at their requirements, we find a number of other substances besides nitrogen and phosphates. These substances are, however, usually present in sufficient quantity in the soil, and therefore the question is generally one of phosphates and of nitrates.

In the extensive experiments at Rothamsted we find investigations upon barley, wheat, oats, turnips, swedes, mangel wurzel, and permanent pasture, covering, in fact, the whole ground that is of direct interest to the bulk of our farmers. In the corn experiments it is interesting to notice the pertinacious manner in which unmanured land keeps throwing up fair crops year after year. It has been pointed out by Dr. Gilbert, that plots of wheat and of barley which have been continuously unmanured for some thirty-five to forty years, are still yielding crops averaging thirteen and fourteen bushels per acre per annum—a yield superior to that of the United States of America, or even of such an old cultivated country as France. If we examine the agricultural statistics of the United States we shall find some States only producing eight bushels per acre of wheat. It is therefore remarkable that these corn-yielding fields at Rothamsted continue for a long period to give such a satisfactory amount as fourteen bushels per acre. But it is not so with root crops. These crops when unmanured give, as you will readily see by examination of the Rothamsted papers, a miserable result. On unmanured plots, that is on plots in which it is attempted to grow roots year after year without any aid, the result is simply *nil*.

Passing in the next place to the effect which is produced by additions of certain of the lesser important substances, I take as examples the application of considerable quantities per acre of sulphate of potash, sulphate of soda, and sulphate of magnesia. The yield is very slightly increased by such applications. In the case of barley it has risen from seventeen and a half bushels upon the continuously unmanured plots to nineteen and a quarter bushels. That is a very small and inadequate increase, and not such as would induce the farmer to incur the expense of supplying such substances.

Superphosphate is, we know, of vast importance, and in

examining the barley plots which have been continuously manured with superphosphate we find the yield forced up to twenty-two and a half bushels per acre, which contrasts remarkably with seventeen and a half bushels upon the unmanured plots. This is the result on barley.

In the case of wheat the superphosphate alone increases the yield from thirteen up to sixteen bushels per acre. Just pause for a moment to consider this important point. Why has the superphosphate acted more distinctly in the case of barley than in the case of wheat? The reason appears to be that the barley occupies the ground for a shorter period, and is a surface feeder, and therefore it is more dependant upon its food being close at hand. In the second place, land intended for barley is reduced to a very fine tilth in the month of March or April, and there is not the slightest doubt that the open nature of the seed-bed for barley causes nitrification to go on in a very brisk manner, and therefore enables the barley to seize upon and to appropriate a larger amount of mineral food. This is not the case with wheat. The ground is close and compact in the case of winter-sown wheat, and the application of superphosphate is probably not assisted by that amount of nitrification which we know is so necessary in order to enable the plant to take up mineral constituents from the soil.

Take, in the next place, mineral applications consisting of all the ordinary mineral materials supposed to be necessary for the growth of plants. They are chiefly in the Rothamsted series represented by sulphate of potash, sulphate of soda, sulphate of magnesia, and superphosphate of lime. It is noteworthy that the effect produced by such applications is but little better than what is obtained by superphosphate alone. Thus you find not only sulphate of potash, sulphate of soda, and sulphate of magnesia produce but little effect over unmanured plots, but likewise that these three substances associated with superphosphate do very little more than superphosphate can do alone.

In the next place, I must call attention to the very marked effects of ammonia dressings applied to the cereals. Ammonia dressings alone, at the moderate rate of 200 lbs. per acre, increase the yield from $17\frac{1}{2}$ up to 30 bushels per acre in the case of barley, and from 15 to 20 and 23 bushels in the case of wheat. I might further support this evidence by taking the case of oats, showing the immense effect of nitrogenous dressings in the form of salts of ammonia on cereal crops. Now, how are the results affected when you add not only ammonia salts, but superphosphate? As our previous reasoning might lead us to expect, the additions of ammonia are rendered much more beneficial when there is likewise added a sufficient amount of mineral plant food. Ammonia alone produces a large amount of chlorophyll in the leaf, and there are symptoms of the plant being over-charged with nitrogen simply because there is an excess of nitrogen in the form of nitrates over the available mineral matter. It is therefore necessary to add mineral matter, and that may be done most efficiently by mixing ammoniacal manures with superphosphates. By this addition the yield rises to 44 bushels per acre, which is not only an immense contrast with the $17\frac{1}{2}$ bushels on the unmanured plots, but is strikingly greater than the $22\frac{1}{2}$ bushels produced by superphosphate alone, and also strikingly greater than the 30 bushels produced by ammonia salts alone. See therefore the importance of giving the plant all it requires, not only plenty of nitrogenous material, but likewise abundance of phosphates.

In the next place I will endeavour to show what benefit occurs if, in addition to ammoniacal manures and superphosphates, you introduce these substances of lesser importance already mentioned, namely, sulphate of potash, sulphate of soda, sulphate of magnesia. That question we are able to answer, for we have plots which are manured not only with ammonia salts, but with sulphate of soda, sulphate of potash, sulphate of magnesia, and superphosphate. Taking the barley experiments, this application gives an average yield of $44\frac{1}{2}$ bushels, or only fractionally more than we obtained from the use of ammonia salts and superphosphate. The same result is obtained when the source of nitrogen is changed from ammonia salts to nitrate of soda in even a more marked manner, for while nitrate of soda and superphosphate have given an average yield of $46\frac{2}{3}$ bushels over a period of 34 years, the addition of sulphates of potash, soda, and magnesia have only given a 47 bushel average. The inefficiency of these substances is equally well shown by the fact that while nitrate of soda has given an average yield of 34 bushels per acre, nitrate of soda plus sulphates of potash, soda, and magnesia has only yielded an average of $34\frac{1}{2}$ bushels. Taking wheat, we see a considerably greater effect. We find $26\frac{1}{4}$ bushels from the use of ammonia salts and superphosphate, and 30 and 31 bushels per acre from the use of various mineral substances, including superphosphate mixed together plus ammonia salts. Still even with the increase in the case of wheat it is evident that the increased yield was rather due to the sulphur or sulphuric acid in combination with the alkalies and magnesia than to the soda, potash, and magnesia. Examination of the tabulated results bear out this idea, because whether sulphate of soda, sulphate of potash, or sulphate of magnesia were used a similar increase is observable, and when all were combined together the effect is not increased at all. It would be difficult to explain the considerable increase from an application of sulphate of soda on any other ground, and the fact that sulphate of potash produced no greater increase than sulphate of soda also indicates that it was not the potash but the combined sulphur that was needed.

Now allow me to point out that in these experiments on wheat and barley the word "silica" does not occur. How is it, I should like to know, that in all examination papers which come into my hands a preponderating importance is attached to silica? There must be some fundamental error at the base of this, for it goes through hundreds and thousands of answers. I am told that silica is the thing to give wheat, and it is silica which is laid stress upon as necessary for corn. It is a very odd thing, and it reminds me of a story by our old friend Dr. Voelcker, in which he related that a farmer exclaimed with reference to an analysis of soil in which some 80 per cent. of insoluble silica was recorded—" My! what a grand thing that inzoluble zilica must be for the ground." I see little in the Rothamsted results as to the effects of silica. I never doubt any ground containing a sufficient supply of silica—of course it does; and there must be some grave error in the teaching of boys all over the country when they rush to silica at once as the most important constituent you have to add to your corn crops.

With reference to turnips and swedes and mangel wurzel, one of the most extraordinary lessons bearing upon practical farming is this—without manure you get no root crop. It is a very significant fact, and bears out what I have already stated, that root crop cultivation is the most difficult, most complicated, and most expensive of all tillages. In turnip cultivation one of the most extraordinary facts is the immediate and rapid effects of superphosphates. Here the deficient material is not nitrogen, but phosphorous. The mineral constituents seem to be wanting in the case of the turnip crop. I cannot help thinking that this is due to the peculiar cultivation of these roots, and the period during which they are growing. Turnip cultivation requires the ground to be thoroughly tilled to a great depth, and to be in a fine condition, the very condition in fact for nitrification to be going on with the fullest possible vigour; and the growth of turnips is carried on through the months of May, June, and onward to October, a period during which the temperature of the soil is suitable for the development of nitric acid. Therefore it appears as if the turnip received its nitrogen from the soil in sufficient quantities, and that the tables are turned; and whereas in the case of corn crops the tension is for nitrogen, in the case of root crops the condition of tension is with reference to mineral matter. I think that partly explains the fact, because turnips require nitrogen. If you look into the composition of a turnip crop you will find it takes up a great deal of nitrogen, but still the application of superphosphates to the soil seems to be especially requisite. Bear in mind also the words used by Dr. Gilbert, that the turnip and root crops are essentially sugar crops, and phosphoric acid seems to be needed where you have crops of this description.

How is it then with the application of these lesser important substances, sulphate of potash, sulphate of soda, and sulphate of magnesia to our turnip crops? The effect is very small. Whereas you get with superphosphate alone $7\frac{1}{2}$ tons, you obtain 8 tons 8 cwt. with the mixture of superphosphate and these other manurial substances. In another case we notice an increase of from 6 tons 1 cwt. to 6 tons 16 cwt. In another case, the third season you get an increase from 4 tons 14 cwt. to 5 tons 9 cwt. only. In another case an increase from 1 ton 18 cwt. to 2 tons 2 cwt., a very slight increase. Superphosphate produces a great effect, but the addition of these other substances gives a small and inadequate but apparently constant effect. The effect of ammonia on the turnip crop is small. For instance, on an unmanured plot I have before me a result of 2 tons 6 cwt., but dressed or crossed with ammonia salts, I see only 3 tons 7 cwt. The effect is not satisfactory, the effect of superphosphate is very great; but the effect of crossing it with ammonia salts in the case of the turnip crop is only to raise the yield from 8 tons 16 cwt. to 9 tons 18 cwt. per acre.

In examining into the effects of various manures upon mangel wurzel we must be prepared for different results. In the first place, we are dealing with a crop belonging to an entirely different order (Chenopodiaceæ), and therefore presumably with a different habit of growth, and selecting its food in a different manner. The striking effect of chloride of sodium upon this crop has been already mentioned, and must be borne in mind in examining the effects of various fertilizers at Rothamsted.

First of all we see brought out strongly in the case of both turnips and mangel wurzel the superiority of farmyard manure. This I have already drawn attention to as in a great measure accounting for the high estimation in which this fertilizer is held. Repetition only confirms the result of previous experiments, and a large number of plots support the conclusion that as far as the root crop is concerned there is nothing like farmyard dung.

Secondly, the need of mangel wurzel for nitrates is abundantly shown. The effect of superphosphate alone is less marked than in the case of turnips, and it evidently requires to be aided by a source of nitric acid, the best being nitrate of soda. It is curious and not easy to explain, that ammonia salts produce a poor effect compared with nitrate of soda except when used upon plots dressed with farmyard manure, when the ammonia salts give equally excellent results. Five hundred and fifty pounds of nitrate of soda, at once send up the yield of mangel wurzel many tons per acre, in many cases trebling and sometimes quadrupling the crop. The effect of superphosphate without nitrate of soda is not nearly so marked as in the case of turnips. The increase over unmanured plots is not striking, but when nitrate of soda is added it appears to develop the action of the superphosphate, and a better result is obtained.

Additions of sulphate of potash, sulphate of magnesia, and chloride of sodium (common salt) often produce a considerable increase on mangel, but it is open to the view that the effect is a good deal owing to the common salt rather than to the magnesia or even the potash. Yet in a crop like mangel, which is a heavy yielder and a rapid grower, a mixture of fertilizers approaching to the composition of a general manure assists in its development. An argument in favour of the view that potash salts and salts of soda and magnesia are not usually required in ordinary practice may be supported by the fact that the mangel plots at Rothamsted are like most of the experiments there conducted, successive cultivations, year after year. A crop such as mangel grown consecutively upon the same land would draw heavily upon the potash and magnesia of the soil—much more so than would mangel grown at the usual intervals in ordinary farming. There would consequently be a probable lack of potash and other mineral ingredients induced. With this consideration in view we may well wonder that dressings of potash salt produced so little effect. We may also reasonably conclude that in combination with a good dressing of farmyard manure they would produce no visible effect, especially as potash, soda, magnesia, and lime abound in farmyard manure.

Next with reference to manuring permanent pastures. The herbage in such pastures is of a very mixed character indeed. It is a very different thing to manure permanent pasture grounds, and to manure a wheat field or even a field of turnips or mangel wurzel. You have plant requirements of all sorts—plants searching after food in different layers of the soil; you have leguminous plants, and plants of gramineous order, and you have also plants of a mixed weedy character, spoken of generically, as weedy or miscellaneous herbage. Now in this mixed population of a permanent pasture, you have, as you might expect, a rather different effect produced by the use of various fertilizing substances; and I may confess that upon permanent pastures you get very extraordinary results by the application of those substances of lesser importance, such as sulphate of potash, sulphate of soda, and sulphate of magnesia, especially when they are assisted with the two better recognized substances—superphosphate and ammonia salts. By far the best yield is obtained by this combination of all the substances which are required. In one case the yield of hay goes up to 4 tons 8 cwt. to the acre, taking both first and second cuttings—that very large yield being derived from the use of sulphates of soda and magnesia, superphosphates, and 600 lbs. of ammonia salts, and for the first time silicate of soda.

Not only is there a very large variety of plants and great variety of requirements, but we must remember that the produce consists not only of grain or of roots, but of the entire stem and flower. There is a very large amount of silica removed from a meadow or pasture soil, and it appears as if this were a case in which the application of silica in a soluble and available form can be made with good effect. In these experiments which have been continued every season from 1856, we find the following average results during eleven years, from 1876 to 1886, taking both first and second crops into the total amounts per acre. The continuously unmanured plots have yielded upon an average 27 cwt. of hay per acre (18 cwt. for the first cut, and 9 cwt. for the second cut); superphosphate of lime at the rate of $3\frac{1}{2}$ cwt. per acre, has only pushed up the yield to $28\frac{2}{3}$ cwt., showing once more the small effect of superphosphate upon gramineous herbage.

Mixed mineral manures composed of sulphate of potash, sulphate of soda, sulphate of magnesia, and superphosphates, show an effect of 44⅛ cwt., which is striking, and well illustrates the effect of mixed mineral pabulum upon a crop like hay, which is cut green, and removes a vast amount of earths and alkalies from the soil. Another similarly treated plot has yielded an average of 48¾ cwt. per acre.

Now look at the effects of nitrogenous dressings alone. The effect of nitrate of soda is (as is also the case in wheat and mangel cultivation) much greater than that of ammonia salts. Two hundred and seventy-five pounds per acre of nitrate of soda applied annually sends the yield up to 40⅞ cwt. per acre, while 400 lbs. of ammonia salts preserves an average of 30 cwt., or but slightly above the superphosphate and the mixed minerals. Two hundred and seventy-five pounds of nitrate of soda with 3½ cwt. of superphosphate raised the yield to 49 cwt., while 400 lbs. of ammonia salts and 3½ cwt. superphosphate only gave 42¼ cwt., still showing the superiority of nitrate of soda for this crop.

The finest results are, however, obtained when a mixture of mineral and nitrogenous manures are applied. The mixture consists of—

>500 lbs. of sulphate of potash
>100 lbs. of sulphate of soda
>100 lbs. of sulphate of magnesia
>392 lbs. of superphosphate
>400 lbs. of silicate of soda
>600 lbs. of ammonia salts

Total   2092 lbs.

This heavy dressing has resulted in an annual average of 88¼ cwt. of hay per acre. That the silicate of soda has an effect is evidenced by the yield of a corresponding plot in which the silicate is omitted, and in that case the yield per acre is only 77½ cwt. The effect of the dressings employed in gradually effecting a complete change in the proportions of the grasses proper, leguminosæ, and the miscellaneous herbage is among the most remarkable results obtained. It is, however, only within my power in the present volume to draw attention to this interesting subject, and to recommend the perusal of the Rothamsted experiments as exceedingly instructive and of great practical importance.

Lastly, with regard to the effect of fertilizers upon such leguminous crops as beans, peas, vetches, and clovers, the conclusion arrived at by Sir John Lawes is, that mineral constituents applied as manure (particularly potash) considerably increased the crops. Ammonia salts produced very little effect, and in some cases have proved pernicious. Nitrate of soda has, however, produced more marked results. When we remember that a leguminous crop contains two, three, or more times as much nitrogen as a cereal crop, it is remarkable that nitrates and ammonia salts should exert little or no effect. The explanation appears to be that they derive their supplies of nitrogen from the deeper layers of the sub-soil, and partly from the air.

The Rothamsted results are worthy of the closest study. They are scattered throughout the long series of the *Journals* of the Royal Agricultural Society. An excellent synopsis of these experiments, so far as they relate to the growth of wheat, barley, and the mixed herbage of grass land, from the pen of Professor Fream (now being published at *The Field* office), will be found exceedingly useful reading, not only for students, but for every one who is interested in this most important subject.

The most notable addition to our sources of phosphoric acid of late years is undoubtedly basic cinder. Basic cinder, or Thomas phosphate powder, is destined to play a very important part as a fertilizer. I have a pamphlet before me in which I am told that no less than 200,000 tons of this cinder are produced in the iron works of Germany, and 130,000 tons are produced in England. It is a bye-product in the system of making Bessemer steel. It is basic in character, being formed from dolomite. The manufacture of basic cinder is carried on by the introduction of calcined dolomite into the molten mass of iron in the converter. As the air, according to the Bessemer system, is forcibly driven through the mass in order to burn off the carbon, it unites with the phosphorous and sulphur, forming phosphoric acid and sulphuric acid, and these acids combine with the lime, and form phosphates and sulphate of lime. It rises to the surface and collects as a slag or cinder, and that is Thomas phosphate, or basic cinder. It is afterwards ground in powerful disintegrators and mills, and is during the process which is progressive exposed to the action of powerful magnets for the purpose of picking out the nodules of metallic iron which it generally contains. This powder is used as a fertilizer. The most extraordinary thing with reference to basic cinder is, that it is most beneficial when applied in the simplest possible form, namely, simply disintegrated. My attention was invited to this subject four years ago, and in conjunction with Dr. Munro I conducted experiments at that time on the invitation of the North Eastern Steel Company, both at Downton and in the County of Durham.

We were unfortunate in 1884. First in selecting mangel wurzel for the purpose of the experiment, a plant less responsive to dressing of phosphatic material than the turnip or swede; and secondly, we were driven to try this experiment on land in a high state of fertility from previous management. The consequence was, our results were, so to speak, negative or neutral, and they did not succeed in showing any great effect from the use of basic cinder. But in the following year we undertook a double series of experiments, partly on Downton College farm, and partly in the County of Durham, on the farm of the Carlton Iron Works Company, and there we got most definite results, which are embodied in the report published by the North-Eastern Steel Company. These experiments established in this country the effect of basic cinder. Other experiments seem to have been going on during the same season in Germany, and both in Germany and in this country the effect of basic cinder as a fertilizer has been thoroughly proved. It is a curious fact that basic cinder should produce as great an effect as superphosphate, which in many cases it does. Undissolved basic cinder produced pretty nearly as much effect as superphosphate with ammonia salts, and a much better result than ground coprolites.

The peculiarity of the basic cinder appears to be that it contains a very large excess of lime in somewhat unstable combination with phosphoric acid. To put the matter briefly, we have in the case of monobasic phosphate one atom of lime in combination with phosphoric acid. In reduced phosphate we have two atoms of lime united with one of phosphoric acid, and in tricalcic phosphate, which is insoluble phosphate, we have three atoms of lime. But at the high temperature at which the basic cinder is formed, we get four parts of lime united with the phosphoric acid. We have a higher proportion of base, and a tetracalcic phosphate produced; but although it is true within limits that the larger the proportion of lime the less soluble is the phosphate, yet the excessive quantity of lime appears to introduce an element of instability into the compound. It readily breaks down, are this appears to be the reason why the basic cinder is more efficacious as a manure than tricalcic phosphate, or even than reduced phosphates.

For the benefit of those who desire to test the comparative value of artificial fertilizers by direct experiment, I will indicate the general lines upon which these experiments should be carried out. First I would select a uniform soil in low rather than in high condition, for example, a soil that had just carried a corn crop, or better still, one that had carried two corn crops in succession. Such soils are in a condition to respond to applications of fertilizers, while the unmanured plots would be likely to yield small crops, which would contrast distinctly with the dressed plots. Secondly, the experiments must be simple, so that we may clearly see the meaning of the result arrived at. We must not ask Nature two questions at once, but a simple, straightforward question. The issue must lie between superphosphate and nothing, or between two hundredweight of superphosphate and four hundredweight of the same fertilizers, or between basic cinder and superphosphate, &c. &c. In order to secure decisive answers, the previous cropping and the previous treatment of all the plots must have been precisely similar. The plots must not extend over soils of two or more sorts, or over land some of which has had an extra cultivation, or yet over soil which has carried more than one and the same crop, or some of which has been dunged. Strict uniformity of conditions must be obtained. Similarly in sowing, hoeing, harrowing, rolling, and any other processes during the growth of the crop, all must be treated alike so as to preserve the simplicity of the issue throughout the whole period. We should always have unmanured plots for the sake of comparison. In experimenting I find it better to have several unmanured plots, and if practicable an unmanured plot alternating with every manured plot. This may be readily done if we arrange the experimental plots in the form of a series of squares each of one-tenth or one-twentieth of an acre. By a proper distribution of unmanured plots we can collect together the average results of our unmanured plots, and we can collect together our average results from the application of superphosphate, ammonia salts, nitrate of soda, basic cinder, or whatever else we may happen to be experimenting upon. Several unmanured plots judiciously placed enable us to discount those rapid changes in the quality of the soil which sometimes take place within a few yards. In a series of experiments where a large number of various fertilizers are tried, it is of great importance not only to have an unmanured plot, but that we should have duplicates and quadruplicates, so that our chain of evidence may be strengthened and brought within the range of strict and immediate comparison instead of a remote comparison with a distant plot on a soil not strictly similar.

CHAPTER XIII.

Rotations of Crops—Early Rotations—Development of—Principles of—The Fallow—Grain and Fodder Crops—Effect of Clover on subsequent Wheat Crops—Similar Effect of Beans—Modifications of the Norfolk Four Course—Catch Crops—Potatoes as a Crop.

THE subject I have next to bring before my readers is that of the rotation of crops. I do not know that I shall be able to say anything very new, but if I can present the subject in its several aspects, and indicate the way in which I think it ought to be taught, that is perhaps all I can expect to do. I remember once lecturing to a class amongst whom was a gentleman who had spent some years in the War Office, and he confided to me towards the end of the term that there was one thing in agriculture he could never understand. I asked him what it was, and he said he could never really comprehend what a rotation of crops was. Of course a rotation of crops is simply an ordinary, regular, and recognized sequence of crops extending over several years; so that there is really nothing difficult to understand about it. But an important question arises on the threshold, namely, how have rotations or successions of crops originated, and also why are they necessary? In answer to the first question, I believe that rotations of crops forced themselves upon agriculturists probably somewhat as follows. When man advances to that stage of civilization in which he ceases to be a mere wanderer, when rights of property become in some measure definite, then he finds that after ground has been cultivated for a certain number of years it ceases to be profitably productive. This we know to be the case when a new country is settled or occupied by immigrants, and when their fields become exhausted they shift their tent further westward or eastward, as the case may be. They forsake the fields which are no longer profitable, and they break new or, as it is sometimes called, virgin soil. In process of time the ground which has been forsaken, or which has ceased to be profitable, regains its fertility in a great measure, and that being the case it may be again tilled, and once more subjected to a series of croppings, which in time again reduce it to a state in which it is unprofitable, and then there is a period of rest—and neglect; so that we have at once a kind of rough and natural rotation of crops oscillating between a period of cropping and a period of rest or neglect. Such a state of things is still extant upon the northern coasts of Scotland. I hear also that it is not uncommon in Norway and Sweden, and I know from personal observation that in Northumberland fields have, even in recent years, been cultivated in this way, cropped as long as they would crop to a profit with wheat or corn, and then allowed to fall away into grazing-ground simply by the invasion of grass and weeds which take possession of the ground. They are then grazed by cattle, and after a few years the land becomes once more capable of bearing a series of crops. That is a rotation in a sense. A period of rest has always been recognized as necessary for land; provision is made for it in the Levitical Law. The Jewish ordinance was that land should be fallowed, or rested, every seven years, and the land was then said to enjoy her Sabbath.

The reasons why land is thus able to recoup itself have occupied our attention previously. It is sufficient, I think, to remind you that those natural forces originally producing the decay and crumbling down of rocks are still in action, and that when we speak of land recovering its energies by rest we

ENGLISH AGRICULTURE.

mean that a certain accumulation both of mineral wealth, and probably nitrogenous and organic matter as well, have taken place owing to the operation of these forces aided by vegetation, and the ground again becomes stocked with available plant-food. That is the natural explanation of the advantage which follows the mere resting of land.

The next point in advance in the history of rotations of crops is the shortening of the period required in order to recoup the energies of land. The systematic fallowing of land is probably very ancient. It was thoroughly understood by the Phœnicians, and it became a part of Roman agriculture, and probably has descended with the succession of civilization from the Phœnicians to the Egyptians, to the Greeks, the Romans, and then to the later European civilizations.

The fallowing of land was definitely introduced into this country by the Romans, and it was introduced from this country into Scotland much later. Even in the earlier years of the last century the systematic working of fallows was unknown in Scotland. The most ancient rotation which I know of is that which once obtained over the whole of Europe, and which is stated by a very competent authority, De Morier, to date back to, at all events, as early as the first century of the Christian era. It is the old Teutonic three-field course, well known in German practice by the ancient village communities of the German Empire, and practised to this very day. It consists in the first place of a period of bare fallow, with tilth and tillage of the land, followed always by winter corn, and that again followed by summer corn; fallow, wheat, oats, or it might be fallow, wheat, beans, or the two might be taken alternately, in which case it would assume the form of fallow, wheat, beans, fallow, wheat, oats. This is the old three-field course which Prof. Rogers in his *History of Agriculture and Prices* states was prevalent over the whole of England in the eleventh century, and it exists in practice in some of our later and less advanced counties at the present time.

The rotation as I have given it is not without its advantages. Perhaps there is no method of cultivating land which enables the cultivator to enter upon his farm with so small an amount of ready money or capital. Two-thirds of the arable land are in grain, and one-third, instead of being cultivated with that expensive plant the turnip, and those plants which are akin to it, is subjected to a system of bare fallowing. Two-thirds in corn and one-third lying idle means a large and rapid return, and when the price of corn was high the return in money was very considerable indeed. Then with reference to the maintenance of live stock, our forefathers and also those nations, such as the Germans, who practised this rotation, always preserved a large amount of their land in permanent pasture.

Every village community of peasants had a considerable proportion of woodland, grazing-ground, and pasture, which formed a ring or boundary of demarcation between them and the next parish, and it was upon this outer ring of pasture-land that the cattle grazed, and from which hay was mown, and the cattle were brought through the winter with the aid of hay, and also with the aid of straw and corn. But at the time when such a rotation as this was practised fresh meat was scarcely to be had during the winter months. Animals were slaughtered towards Martinmas, they were salted down, and the population fed upon bread and salt meat during the winter months, to the great disparagement and injury of their health, and the extension of a large number of diseases which happily have now disappeared, such as scurvy and leprosy.

The next point in advance was undoubtedly the introduction of fallow crops, doing away with the bare fallow, and substituting for it root crops and various crops akin to root crops. I may remind you of the reasons why these root crops may be properly substituted for the bare fallow. I have insisted on this on a previous occasion, but there will be no harm in again mentioning that the root crop is essentially suited for taking the place of the old bare fallow. The root crop is consumed upon the premises, therefore it does not remove any important constituent, as is the case when wheat or other grain or products are removed from a farm. In the second place, the late period at which the root crop is sown, coupled with the care which is taken during the entire growth of the crop, insures the thorough cleaning of the ground. Then again the root crop especially requires the ground to be in a state of high fertility. It is a crop which more than any other suffers from the want of abundant nutrient matter, and therefore the ground must be thoroughly well dunged, and well assisted by other aids to fertility. So that in the root crop we have really all that is required in a renovating crop. The ground is also cleaned when under a root crop; it is thoroughly tilled, and the root crop leaves the ground well stocked and stored with available plant-food chiefly, and in fact altogether, because it is eaten upon the land or upon the farm. It may as well be pointed out that such would not be the case if the root crop were removed or sold. Remove or sell the root crop, and we should find the exhaustion of the ground would be very much more rapid and very much more severe than by the selling of corn crops. The root crop, then, is not renovating except from the fact that it is consumed upon the farm where it is produced.

In all rotations of crops the fallow or fallowing crop ought to take the first place, because it is during the fallowing process that land is prepared for supporting a series of crops. Therefore in enunciating or teaching a rotation it is very much the more convenient method to deal with it as the first of the series—as the very foundation for cropping.

The second year of the rotation will almost always be devoted to grain, the kind of grain depending on the nature of the soil. The grain crop will be followed by a fodder crop for the third crop. Then there will be another grain crop during the fourth year, and it is possible that we may then have recourse to the fallow, in which case we have a four-course rotation, the fallow commencing a fresh series or new rotation. In other cases the succession of crops proceeds further. For example, the Norfolk rotation is a familiar instance of what we have just been considering, that is, roots followed with barley, followed with clover, followed with wheat, followed again with roots as the commencement of a fresh series or rotation. What is sometimes called the Holderness rotation is an example of a more prolonged course, as, for example, when roots or fallow are followed with wheat, followed with clover, followed with wheat, followed with beans, followed with wheat, which is a six years' rotation. The period which elapses between the fallowings limits the length of the rotation. Upon poor land, fallows must be frequent, especially where a somewhat exhaustive system of cropping is pursued. Thus we have the three-field course, that is, fallows every third year—fallow, wheat, beans. Upon a large number of soils the four-course rotation may be adopted, and on richer soils a six years' rotation, as already explained. Upon fertile clays we may go a step further, and place a seven or eight years' interval between the periods of fallowing.

It is convenient to separate between rotations suitable for heavy lands, and rotations which are suitable for light soils. Most of our soils may be conveniently classed in one of these divisions, either as heavy or light. In planning rotations for heavy lands we ought always to be careful to select crops which affect heavy lands. All kinds of land will grow all kinds of crops, but there is a suitability, and in our capacity as teachers or examiners we ought to encourage a student to be definite. If he is asked for certain information, let him give it definitely in the line that is asked for, and not confuse his answer by introducing matter which may be according to good practice in some districts, but at the same time is not according to the principles which you might be supposed to insist upon in teaching the best practice. Therefore, in prescribing rotations for stiff land let us stick to stiff land crops.

I will give some further illustrations of what I mean by going through the three, four, or more years which constitute a rotation, the first of which is, as I have already stated, the fallowing year. Now if there is any position at all in which a bare fallow may still be recommended, it is upon stiff land. Therefore in planning a rotation for stiff land I should consider it right to recommend a bare fallow, with qualifications, because although bare fallowing may be properly used upon stiff land, yet there is a class of crops, root crops and kindred crops, which are yearly encroaching upon the area of bare fallow. Besides that, thorough drainage, the application of steam for tillage purposes, and the introduction of improved implements and fertilizers, have placed clay land more within our power than formerly, and what with these improved appliances and the introduction of a larger variety of crops, the domain of bare fallow is a diminishing area. Upon a really stiff land I should say a bare fallow might occupy a position, but also cabbages might be cultivated with great advantage, as well as all those crops which are akin to cabbages, such as thousand-headed kale, brocoli, curly kale, kohl-rabi, and rape. All of these plants are thoroughly adapted for clay land. They have strong, penetrating, deep-searching roots, and they do not require the ground to be in such a fine state of tilth as some of the other root crops. Rape especially may be drilled upon clay land, which would scarcely be considered fine enough for turnip cultivation, and it resists the turnip-fly. It is a capital crop for strong land. Rape also is fed off at a period of the year when clay land will bear sheep. It is not entirely for winter feeding. Rape may be sown so as to be ready for feeding in July, August, September, and October, and during all those months the land would not suffer from the treading of sheep. It is a capital preparation for wheat. I do not know any crop that wheat is so likely to thrive well after as rape. Again, some of these crops are best raised in seed-beds and transplanted out into their positions in the fields, and this is a great relief to clay land farmers, as may be shown by illustration. You have already heard that clay land works most mellowly and pleasantly immediately after harvest; after it has had the summer sun upon it clay land will often work in an extraordinarily friable manner, so that no one would believe it was the same leathery, compact, "planky" stuff that ploughs up in the month of March. There is a period after harvest during which clay land can be wrought into a very fine condition. I always like to have my cabbages and kale sown in a suitable position the first week in August. They are sown on a finely-wrought half-acre of ground, and they very quickly make their appearance. While the young cabbages are growing, the ground in the fields where the crop is to be planted is thoroughly worked, dunged and ridged up, and brought into the right condition for the reception of the cabbage-plants. This is a cultivation which is suitable for clay land. The cabbages stand the winter; they are hoed, and horse-hoed, and top-dressed in the spring, and will be ready for feeding in July, August, and September—a period which is well suited for clay land. The same remark applies to kales; thousand-headed kale is a capital crop, where the situation is favourable, but many good farmers now prefer to sow it in seed-beds and transplant it out. Observe what an opportunity it gives the clay land farmer if he has his seed raised in a seed-bed, and then transplanted out after properly tilling and working his clay fields. The same remark applies to kohl-rabi, which is another excellent fodder crop, we cannot call it a root crop. Raise kohl rabi in a seed-bed, and plant it out in May, June, and July, or even in August. The May plants will give an autumn feed, and the July and August planting would give a spring feed. Then again mangel wurzel, although it dislikes the very stiffest class of clay, yet it does well on clays which are rather too stiff and hard for turnip cultivation. Mangel wurzel is a most excellent crop for the South of England, or for a dry summer, and it does best upon lands which incline rather to the sort of which I am speaking. We may also to some extent encroach on our bare fallow, especially on the lighter portions of it, by sowing a few swedes and white turnips; so that, taking it altogether, there is an opportunity of placing a good deal of our heavy land under some sort of fallowing crop.

The Holderness rotation is adapted for stiff soils of good quality, and is usually rendered as follows:—

    1st year. Fallow (bare or cropped).
    2nd   „   Wheat.
    3rd   „   Clover (mown or fed).
    4th   „   Wheat.
    5th   „   Beans.
    6th   „   Wheat (previously dunged on bean stubble).

This rotation well exemplifies the principle already laid down. It is an alternation of fodder or fallow crops with grain crops. If we consider that the beans are grown for home consumption, they may be fairly looked upon as a fallowing crop. Beans require careful hoeing and inter-culture, and if properly cultivated may be regarded as a cleaning crop; and if we allow them to be so regarded, then the above rotation may be said to consist in an alternation of fallow or fodder crops with grain crops.

It is necessary to discriminate carefully between rotations suitable for the two great classes of land known as light and heavy lands. We must not forget that there are many descriptions of soils, each of which is specially adapted for certain systems of farming. The two chief classes of soils are, however, heavy or clay soils, and light or free working soils. I have already drawn attention to the entirely different methods in which tillage operations are carried out upon these two classes of soils. In order to understand the principles upon which they are cropped it will be necessary to take them separately, and examine rotations which are found suitable to each.

Both are cropped alternately with fallow and grain, or fodder and grain, but the kind of fodder crop or grain crop introduced is a point which requires our best attention.

Taking the case of clay land, the first year of the rotation will be devoted to fallow. If the land is foul, naturally wet, or if the season is late and damp, it is probable that a portion at least will be treated as a bare fallow. If, on the other hand, the land is well drained, clean, and the season be favourable, we shall endeavour to cultivate it for root or fodder crops. We must remember that we are dealing with wheat land, and our root or fodder crops must be selected with a view to following them up with wheat. Thus rape, cabbages, early sown white turnips, early sown swedes, and mangel wurzel may all be appropriately grown on clay soils, as they allow of being consumed in the late summer month when the ground will not suffer either from the treading of sheep or the passage of carts over its surface. Vetches, also, may be grown upon the fallow portion of clay land for summer consumption by sheep, or for soiling at home, but cannot compare with the crops already mentioned for cleaning purposes. If vetches are taken, it will be necessary to half-fallow or rag-fallow the land after they are disposed of, with a view to getting it ready for wheat-sowing in September. We have heard much of wheat being grown at a loss; but as long as stiff clay land is retained in arable cultivation, so long will wheat be found the most suitable and most paying of the cereals upon that particular description of land in all localities where the climate is suitable for it. If it does not pay, we may be sure that no other corn crop will prove remunerative. The second year of the rotation will then be devoted to wheat. The third year we shall have clover, or mixed clover and grass seeds, and if the land is in good condition we may expect a heavy cut, followed by an abundant feed for sheep or cattle. Clover is one of the best preparations for wheat. It is furnished with roots which penetrate the subsoil to a depth of several feet. At Rothamsted land under clover was found to have been robbed of its nitrates to some fifty-four inches of vertical section. These nitrates are brought up to the surface. Simultaneously it removes carbon, and possibly nitrogen, from the air, and then fertilizing matters are made available for succeeding crops. The fact that a mown crop of clover leaves the land in better condition than a fed or grazed crop at first sight seems almost paradoxical. It is, however, easy of explanation, when we remember that a mown crop of clover has been allowed the opportunity of full development; whereas a grazed crop has not been able to rear its head upward, or to strike its roots downwards, as in the case of the mown crop. It has been perpetually snubbed and checked. The fully developed plant, fit for mowing, and in full inflorescence, has done its work thoroughly, and has absorbed its due weight of solid substance from the subsoil and atmosphere. Its roots have reached their full development, and the fall of leaf during the continuous stages of its growth have top-dressed the surface with rich, fertilizing material. It is these considerations which explain why a fully developed and large crop of clover, even when cut and carried away as hay, leaves a field in better condition than does a constantly nibbled crop of clover.

The fourth year of the rotation on stiff soils will again probably be wheat. We cannot do better. The ground is in good condition for wheat, and is naturally suited for it, and therefore into wheat it must go.

The fifth year, according to the Holderness rotation, we take beans. Beans are well adapted for clay land—such soils often being designated as "wheat and bean lands." Beans also form a very excellent preparation for wheat, acting in a manner somewhat similar to what has already been found in the case of clover. They too pump up nitrates from the lower layers of the subsoil, and absorb solid matter from the air; and they too drop their leaves upon the surface soil during the period of their growth and maturation. The sixth year of the rotation accordingly is devoted again to wheat. Thus, in the course we have proposed, half of the land is under wheat, and two-thirds is under wheat and beans, the remaining third being devoted to fallow and fodder crops.

The four-course rotation may be adapted to stiff land by devoting the first year to root crops suitable for heavy lands, and taking wheat instead of barley the second year. Barley, it is true, may be, and often is, grown upon clay land, but the quality is inferior, and the grain is not suitable for malting purposes. It may be ground up for pigs, or sold for grinding or distilling, but the crop is at a disadvantage when grown on clay soils. Wheat, on the other hand, is always a crop upon which we may rely to give a fair return upon retentive clay soils. Wheat may be taken a second time in the rotation after the clover, the rotation then reading—roots, wheat, clover, wheat; or oats may be substituted in the last year of the rotation.

The old-fashioned three-field course already noticed may be adapted to modern requirements without altering its general principle. The fallow may be cropped to a greater or less degree, according to the character of the season and the condition of the land. If, for example, we take suitable crops, such as cabbages, rape, mangel, or even a few swedes and early sown turnips for late summer consumption upon half of the fallow breadth; and if we divide the portion usually devoted to beans, and take instead, half beans and half clover, our rotation will then read as follows:—

      1st year.   Fallow (bare and cropped).
      2nd „   Wheat.
      3rd „   Beans or clover.

And if the portion which has been in beans is brought into bare fallow, and the portion in clover is brought into cropped fallow, then the rotation may be written out as a six years' course:—

      1st year.   Fallow.
      2nd „   Wheat.
      3rd „   Clover.
      4th „   Swedes, &c.
      5th „   Wheat.
      6th „   Beans.

With regard to rotations for light lands, the four-course or Norfolk rotation is the best known and most widely practised. It is so familiar to most students of agriculture that it seems unnecessary to repeat it, but with a view to its many modifications it is as well to state it in its simplest and best known form of—turnips, barley, clover, wheat. This rotation is evidently suitable for the lighter classes of soils. The first year is devoted to turnips or swedes, which are usually consumed upon the field that grew them by sheep. Light land will grow heavy crops of turnips, and such soils derive great benefit from the treading of sheep in the winter. In ordinary good farming the turnip crop is not fed alone, but with fair allowances of hay, oil-cakes, and corn. This fact is often lost sight of by candidates, who seem to think that the benefit of folding is really a balance between the material of the crop removed by the animals in increasing their live weight, and the remainder of the crop which finds its way into the soil in the form of animal excrements. This is, however, a practical error, as the benefit to the land is really expressed by the total manurial value of the entire crop consumed, *minus* the amount retained by the animal for its own increase, *plus* the value of excrements derived from imported foods—which last is a most important item. Another advantage to the soil in contradistinction to the subsoil is the importation to the surface soil of material collected by the root crop from the subsoil, to which may well be added material from the air. Add to these effects the improvement in texture to sandy or light calcareous soils, and we have the sum of the benefits which accrue from a system of sheep farming on arable soils. Barley is decidedly the most suitable crop for the second year, as in these days a better value is obtained upon suitable soils from barley than from any other cereal. On the other hand, it may be remembered with advantage that it is easy to "overdo" land intended for barley, and that the best samples of malting barley are not produced on land too highly fertilized. Clover naturally takes up the third year, and for reasons already given, as well as the firmness which its growth insures, is an excellent preparation for wheat.

The rotation, therefore, is well drawn, and truly scientific in its object and arrangement. It is not without its faults, but those faults are happily of a nature which can be readily corrected. The shortness and the paucity of variety in the crops embraced within the rotation as above given are the most serious drawbacks. I have already pointed out the possibility of substituting wheat for barley in the second year as suitable for clay lands, and it is equally feasible to take oats instead of wheat in the fourth year. These changes are not destructive of the principles of the rotation, which still reads—roots, grain, fodder, grain—as before.

Another more important modification—simply effected—is to allow the seeds to remain down an additional year, thereby changing the rotation from a four-course into a five-course rotation. Simple as this is, it involves considerable changes in the economy of a farm, and these changes are still more marked when the suggested alteration is carried further by allowing the seeds (clover) to lie three years, or, as sometimes is the case, four or more years. Such deviations from the Norfolk rotation are more popular in the north and west than in the south-east of England. They are a step in the direction of laying land down in pasture, and point to a larger head of live stock and lower acreage of corn. On the strict four-course system, one-half the arable land is in corn, and one-half is in root and fodder crops; but on the five or six-course two-fifths and one-third are respectively under corn, and three-fifths and two-thirds are respectively under crops suited for live stock. Among other changes, the labour bill is reduced, the number of horses can be lessened, and seed bills, manure bills, and other charges, are also proportionally diminished. In all districts where there is an abundant rainfall these changes are beneficial, but they have never found favour with farmers in the drier counties, such as Norfolk, Suffolk, Essex, and Kent. In the corn-growing districts the climate is better adapted for cereals than for grazing, and it is therefore found advisable to keep up the corn area. Another objection often urged against allowing clover to lie two or more years is the encouragement the system affords to wire-worm, which becomes a serious plague when it once takes possession of a field.

The Northumberland rotation is merely a modification of

the Norfolk four-course, consisting in allowing the seeds to lie two years, and taking oats as the fifth or last crop. North country farmers also find that wheat succeeds better after turnips than does barley. This appears to be owing to the fact that barley cannot be grown of the same high quality in the north as it can be brought to in the south. Secondly, that wheat does not thrive well after clover or "lea" in the north, probably owing to the colder climate, which prevents the rapid decay and nitrification of the clover root, and its passage into the condition of plant food. The Northumberland rotation reads as follows: Turnips, wheat, seeds, seeds, oats. The advantages of these changes are very great; first, in securing the benefits already claimed for a preponderance of grazing in a cool and moist climate; and secondly, in producing in the oat crop a valuable fodder both in the form of corn and straw. By this change it is practicable to let the seeds of clover lie much longer before ploughing them up than when wheat is to be sown. Land for oats may lie unploughed until January or February, giving the occupier the entire benefit of the autumn and winter feed, which is considerable when a large flock is kept. Besides this, there are subsidiary advantages in the supply of oat straw, which yields better fodder than wheat straw; and a further advantage is gained in the greater ease and less expense of ploughing the lands in the winter months.

The four-course rotation may again be altered by increasing the variety of its crops. Turnips may be grown too frequently, the consequence being anbury, clubbing, and finger-and-toe. There is no objection to vary the root-crop within its ample limits by substituting, according to the nature of the soil, swedes, rape, kohl rabi, cabbage, kale, or mangel, thus relieving the ground from the rigid system indicated by the formula—turnips, barley, clover, and wheat. Wheat or oats may be substituted for barley, according to circumstances of soil or season. The third year is perhaps the weakest part of the rotation, as it is known that clover cannot be successfully grown at such short intervals as four years. This disability chiefly belongs to red clover (*trifolium pratense*), and a ready means of evading clover sickness is to omit red clover from mixtures of grass and clover-seeds, so that the plant shall be sown only once in eight or twelve years. Another alternative is to take beans or peas instead of clover, and a third is to grow Italian rye-grass instead of clover. A rotation thus modified will appear somewhat as follows, each crop being considered as representing one year:—Turnips, barley, *clover*, wheat, swedes, oats, *peas*, barley, cabbages, wheat, *Italian rye-grass*, wheat.

The introduction of fodder crops as "catch crops" before "roots" is another important modification. It is practised to some extent in most counties, but in none more than in Wilts and Hants, where it is co-extensive with the entire area under root crops. The system consists in preparing the wheat-stubble intended for roots and sowing it with trifolium, or with rye, barley, or vetches. As these are fed off by the sheep in the spring, the folds are broken, the ground is reduced to a firm state, and mangel, turnips, or swedes are drilled. The result is a system of double cropping on the fallow breadth, well calculated for the maintenance of a heavy sheep stock. Under this system good root crops may be grown in the southern counties, where these crops can be sown later than in the north. Owing to climatic peculiarities, it could not be carried out successfully in the northern counties. It is essentially a system for free or easy working land, and for a southern climate.

On all arable farms, where large flocks of sheep are kept through the winter, there is a difficulty in managing the later feeding of turnips or swedes on the land. The season is waxing late, and it is time that the sowing of spring corn should be brought to a conclusion. Still, the sheep block the way, and when the spring is late we may easily require to fold turnips until the middle of May. In such cases the question arises as to what must be done with our late root land, which according to the Norfolk rotation would naturally come in to late barley. It is here that the Wiltshire rotation comes to our relief. This rotation is much practised in the chalk districts of Hants and Wilts, occupied by the large and well kept flocks of the improved Hampshire Down sheep. It consists in alternating the usual Norfolk four-course with a rotation in which two root crops are taken in succession, followed by two corn crops. It varies according to circumstances, but may be expressed as follows. First year, winter rye, or other early fodder crop, fed off with sheep, the land being then broken up for a full crop of turnips and swedes, also to be fed upon the land; second year, barley; third year, clover and seeds; fourth year, wheat; fifth year, winter vetches, fed and followed by a late crop of turnips; sixth year (after feeding off the late turnips), early turnips, also fed off by sheep; seventh year, wheat; eighth year, barley.

This system may be varied, as for instance by allowing the seeds to remain two years, which would convert the rotation into a nine, or it might be stretched into a ten, years' course. Where adopted it is found to meet the difficulty above pointed out with regard to the latest fed turnips, as these are naturally selected as the site for taking the second turnip crop. It is also a matter of experience that a better sample of malting barley can be grown after wheat, as above indicated, than can be grown immediately after a root crop eaten on the land, with addition of hay and purchased foods.

These rotations may be also varied by the introduction from time to time of sainfoin, which may occupy the ground for from two to six or seven years, thereby giving it a complete change. As remarked in another place, sainfoin ought

ENGLISH AGRICULTURE.

not to be grown on the same land more frequently than once in twenty years.

Upon rich alluvial soils, the heavier loams, and the soils of the old and new red sandstones, potatoes often occupy an important position in a rotation of crops. Thus near Dundee, on the rich soils of the old red sandstone, a rotation is practised known as the East Lothian system, consisting of a six years' course arranged as follows:—

  1st year. Turnips and swedes.
  2nd „ Barley (half dunged).
  3rd „ Clover.
  4th „ Oats (top dressed).
  5th „ Potatoes.
  6th „ Wheat.

Potatoes, although in a sense a fallowing crop, or a crop which insures land being highly manured, clean, and well worked, is still an exhausting crop, because it is sold off the farm. In the above rotation they occupy a position not dissimilar from that of beans in the Holderness rotation, but take the place of a corn crop rather than of a fodder crop. The objection to potatoes superseding the genuine root crops is that every acre devoted to potatoes directly infringes upon the area set aside for live stock, and if cultivated in this position, on an extensive scale, would cause a necessary diminution in the number of sheep and cattle maintained.

CHAPTER XIV.

Depreciated Value of Clay Lands—How to maintain Live Stock on Clay Soils—Catch Cropping a Matter of Situation—Soil and Climate—The Theory of Rotations—Clover Sickness—Practical Advantages of Rotations—Purifying Effect on Land for Sheep—Advantages of Light Soil—Laying Land down to Grass—Its Difficulties—Its Expense—Its Tediousness—How best to Bridge over these Difficulties.

OF all classes of soils, the heavy lands have suffered the most during the great depression which hangs over British agriculture, and the reasons for this are first, because corn is the staple production of arable land of clayey character. We know that the value of corn crops has gone down to an extent which must be looked upon as a great national misfortune. It is not many years since good farmers on clay lands could look to the realization of £12 to £13 per acre on their corn crops, and if we were to place the figure at £7 or £8 per acre at the present time we should be quite as near the actual value. While there has been this tremendous depression in the value of the staple product of clay land there has been no adequate diminution in the cost of production. In the next place, clay lands are critical to work, and in the run of wet seasons, which we may consider as now ended, but which ushered in the unfortunate decadence of agriculture now apparent, clay lands suffered greatly from want of proper cultivation. Another reason why clay land farmers have suffered has been that they have not been able to enter into successful competition with the farmers of light land in respect of live stock breeding, and especially in the production of mutton and wool. Many efforts have been made to improve these unfortunate conditions with respect to clay soils, and among those a rotation of crops has been proposed to enable the clay land farmer, while growing corn, to compete in the growth of live stock, and the production especially of mutton and of wool. The following rotation is worthy of consideration. It commences with winter vetches fed off, let us say, in the month of June, in good time to be broken up and sown with white turnips, rape, or even a second sowing of vetches. These second crops are ready to feed off in September, before the winter's rainfall, and the ground is then in good condition to be ploughed up and sown with wheat. The wheat ground is seeded down half with trefoil, which is a very early crop, and half with mixed seeds. That brings us to the third year. During the third year the trefoil is mown in the month of May, and it is then broken up for rape or turnips. The second portion is mown and fed. During the fourth year the whole of the share comes in for wheat once more, and in the fifth year all goes into winter beans, an excellent crop for clay lands, and the rotation is re-commenced by breaking up the bean stubble and getting it into winter vetches.

Notice with reference to this rotation that it is well contrived and scientifically framed for the accomplishment of certain objects. Three-fifths of the entire arable land is under corn, and not only that, but under descriptions of corn which are especially suitable for clayey ground, wheat and beans; the remaining two-fifths are under crops which are thoroughly suitable for sheep; and further, these crops are all consumed during the summer, at a period of the year when clay land will not be injured in the least degree by treading. I have known this system, or a modification of it, carried out by an extensive Wiltshire farmer, who occupied both heavy and light land. By a system similar to this he was able to support his sheep during the summer months upon his clay land, and during the winter months they went on to his lighter land. The system is, however, not free from objection, the principal being that there is too large a proportion of work thrown upon a particular season of the year. When we reflect upon the apportionment of labour we cannot help noticing that such a large proportion of winter vetches, winter wheat, and winter beans must be a heavy tax upon the horse-power of the farm. The whole of these crops depend in a great measure on their being got in in good time. It is no use sowing either wheat or winter vetches too late, especially upon cold lands. To have all this important work thrown upon us during the short period between the end of harvest, and, let us say, the middle of November, forms, I think, rather a solid objection to carrying out this rotation in its entirety. It does not, however, militate against the principle upon which the rotation is constructed, and if the principle is correct, it may be adopted to a certain extent, if not over the entire area of a large farm.

There is yet another objection to it of a practical character, viz. that in the northern part of England and Scotland it would be difficult to grow vetches, or even trifolium, in conjunction with a succeeding turnip crop. The seasons are later, and vetches would therefore not be ready, while trifolium would be liable to the same objection. A further difficulty arises from the fact that the exigencies of our northern climate seem to require that the root crop should be sown early. Just look at this fact for a moment. In the North of England we always begin swede sowing on the 13th of May, and enterprising farmers—I do not say they are right—would sometimes get their swedes in before the end of April. Again, it is an axiom in that part of

Northumberland that it is not wise to sow white turnips after about the 20th of June for a good crop; and it is just probable that, as we go further north, we should find that the tendency is to plant the turnips and swedes earlier still. At all events, the tendency is towards later sowing as we come south. In the southern counties of England we do not think of sowing swedes until about May 25th. We consider that we are all in good time, and that it is the very prime of the season, if we can get our swedes sown during the first week of June, and those who are not fortunate in getting their crops sown so early continue sowing right through June and into July. Yellow turnips may be sown up to, at all events, the middle of July, and white turnips, especially late white turnips, may be sown all through July, and up to or about August 12th. Later than this we consider that it is a lottery whether turnips come to anything or not. Notice how conveniently all this fits in with a system of catch cropping or fodder cropping, previously to turnips, in the South of England. But it does not fit in so well in the North.

Before leaving the subject of rotations of crops I must say a few words more upon their theory. It has been pointed out that rotations are forced upon us; but when we look a little closely into the matter we shall find that the reason why rotations are important is that certain plants may very readily be grown too frequently upon the same land. Take, for example, red clover. When the plant is grown at intervals of four years, or even longer, the land becomes clover sick. It will therefore never do to attempt to grow clover repeatedly upon the same ground. Sainfoin may be used as another example. Sainfoin requires a much longer period. It is not easy to understand, but it is a fact, that sainfoin requires something like twenty years to elapse before it can be successfully cultivated again on the same land, at least that is our experience. Some people say once in a lifetime; other people say once in a lease. Nature clearly points out in the case of sainfoin the necessity of change and of some form of rotation of crops. The same rule holds good with flax; much land will grow flax, but we must not grow this crop frequently, perhaps once in twelve years will be found sufficiently often. Clover seeds cannot be grown often on the same land. I could hardly recommend you to grow clover seeds more than once in twelve years. Potatoes are much the same. In field cultivation land will often grow one good crop of potatoes, a bumper crop, and it is in vain that we try to grow another, especially on land of rather a weak character. Other examples could be given, but these will suffice. Cereal crops are not open to this objection; they can be grown continuously on the same land.

When I was first introduced to the scientific aspect of agriculture, the question of clover sickness was a very important one, and there was a great deal of controversy as to the reasons why land becomes clover sick. I find the same controversy going on now, and I am not aware that it has made much advance. Want of food is perhaps the reason, but why that should be the cause it is very difficult to understand, because, after all, the requirements of clover, according to the chemical analysis both of the ash and of the plant itself, ought, we might imagine, to indicate the kind of food required, and that kind of food could be added to the land. But all efforts to do so up to the present time have been failures. The Rothamsted experiments with regard to clover sickness will well illustrate this point. In those experiments the land had not only been liberally and completely manured for the first nine inches in depth, but it had been manured for the second, third, and fourth nine inches in depth successively and simultaneously. The ground having been manured by special means down to three, four, or five successive sections of nine inches each, still refused to grow red clover. There is a standing instance in the garden at Rothamsted in which red clover has been continuously grown for about thirty years without any trouble at all. The ground upon which this successful experiment is proceeding having been for a period of three hundred years a kitchen garden, is therefore no doubt thoroughly well stocked, as in fact was proved by analysis, with both nitrogenous matter and available mineral constituents. Here there is no clover sickness, and the reasoning seems to be in favour of the fact that clover growth is a question of food supply. It appears, however, practically impossible in ordinary farming to keep a soil in a sufficiently rich state to the required depth, neither can we have the constituents of the clover food in a proper condition for securing a continuous growth of clover upon arable land for a series of years. Probably a similar explanation is the most reasonable one as to why we cannot grow sainfoin or flax for an indefinite number of times—the food is wanting. It has been held that crops may render the ground unwholesome, or in some way unfit for the growth of similar crops. This view was promulgated under the name of the excretory theory, or, in other words, that there were certain emanations or excretions from plants which in some way rendered the soil distasteful to crops of the same kind, but which might render it all the more fit for the growth of other crops. Unfortunately no proof whatever has ever been vouchsafed as to any emanation or excretion sent out by the roots of plants. The long period which must elapse between certain crops therefore appears to be due to a fact which is not thoroughly explained, namely, that certain crops render the soil unsuitable for their own description of crop for a considerable period of time.

A second reason for rotation is that plants feed differently.

An inspection of a table of analysis of plants indicates a general resemblance in the constituents they require. They perhaps do not take up the substances in the same proportions, but there is a fair general resemblance in the composition of both the soft tissues and the ash of plants. But there is a very great difference in root distribution, so that we must look for the explanation of the value of the rotation of crops partly in certain facts which belong to chemistry, that is, the exhaustion of certain available materials required for a particular plant; but we must also look at the question from the physiological point of view, and remember that the root distribution, *i.e.* the section of soil in which the root labours, and the amount of absorbent surface which the root is able to present to the soil, has a great deal to do with the advantages of a rotation of crops. To grow wheat or any other crop year after year upon the same land, thereby removing from the soil identically the same substances, and exposing it to exactly the same root distribution, taking nourishment from the same section, must be an economical mistake. It would be a flagrant blunder, even if carried out with a commercial profit. Much wiser would it be to interpolate a crop of beans, or to take a crop of clover, in which case you would introduce a plant which asks not only for a different amount of plant food, or a different proportion of plant food, but which likewise seeks that nourishment in quite different layers and sections of the soil.

These are sound reasons why a rotation of crops is in every respect to be recommended, and there are certain subsidiary reasons which are less direct, but which are at least to be taken into account. A proper rotation of crops undoubtedly ensures the proper cleaning of the ground. It is exceedingly difficult to keep land clean under continuous corn growing, and they find this a serious item of expense at

Woburn and at Rothamsted, where they carry out a system of consecutive corn-growing for experimental purposes. I cannot believe it is done at a profit. We know perfectly well that the Woburn experiments have been a heavy tax on the liberality of the Duke of Bedford, and that the experiments are carried out at a great expense at both places, chiefly because of the vast amount of trouble that is requisite in order to keep the ground clean; whereas, if we take an alternate system of cropping we have a natural and easy method of cleaning the ground. That ought not to be lost sight of. It certainly ought to be impressed upon all students that the cleaning of the land is one great reason for an orderly succession of crops. Another collateral advantage is, that rotations preserve a uniform amount of work during the whole of the twelve months. All corn growing or all root growing would throw a vast amount of labour on to particular seasons of the year, and leave the farmer and his men idle for the remainder. A constant staff of men and a regular force of horses may be looked upon as one of the advantages of a proper rotation of crops; and lastly, it is only by a systematic rotation of crops that we are able to support our sheep throughout the entire year without sudden variation in number. It would not suit a farmer to be subjected first to a period in which he had an excessive lot of keep, necessitating his going into the market to buy, and then another period in which he would be compelled to go into the market and sell. He wants to so crop his land that he has a provision for live stock during both the winter, the summer, the spring, and the autumn.

Another reason why a rotation of crops is of great value indeed, especially with reference to this last point of live stock, is as follows. If we were to apportion our land on the principle that this is a wheat-field, therefore let us grow wheat always upon it; this a field suitable for turnips, therefore let us grow turnips upon it year by year, what would we experience? We would find among other disadvantages that our land devoted to stock would become "stained" and foul, not from weeds, but it would become stained and unwholesome for sheep. We cannot carry sheep over a field several times with close folding without noticing that the sheep are not thriving. We can see it in the wool; and in their whole appearance. We notice it in the fact that they "scour," or are subject to diarrhœa, and above all we notice an increased mortality. The corn crop is the great purifier of arable land for sheep-farming, and if there were no other reason than this, we would find sheep farmers would still be compelled to carry out corn cultivation. Every shepherd knows it; the ground becomes sour and unwholesome, but a wheat crop or a corn crop seems to remove everything which is deleterious, and it comes out again fresh and good for the feeding of sheep upon it. I look upon that alone as a very great reason, and one seldom insisted upon, for the maintenance of rotation of crops.

The next point I wish to direct attention to is the question of relinquishing crop cultivation, and letting the land once more return to a natural condition—that of permanent pasture; that is, in other words, the question of the abandonment of agriculture altogether, and betaking us to the semi-barbarous occupation of pastoral life. It is rather a sad alternative in a scientific age like this, but it is the tendency, as we see by an examination of agricultural statistics. I believe there is about one-sixth less corn grown in this country now than there was twenty years ago, and the amount of permanent pasture has increased in proportion. There are certain classes of land in which this change is progressing, and there are other sections of land in which arable cultivation is still likely to hold its own. On all stiff clay soils there is a strong tendency to let them fall back to grass, partly on account of the difficulties which beset the cultivation of clay land, which I need not repeat, partly also because of the special adaptability of these grounds for grass. On these soils we are deterred from arable cultivation on the one side by difficulties, and are encouraged to lay them away to grass by the great relief it offers, and by the fact that they are thoroughly well adapted for the purpose. On the other hand, with reference to light lands, if they can be taken at a sufficiently low rent,—they can be worked at comparatively light expense,—they are safe cropping soils, and their products have always commanded a better price. Sheep have been profitable for the purpose of agriculture for very many years, and sheep are still a hopeful kind of stock so far as profit is concerned. With sheep we must take into account wool. Then, again, barley has many points to recommend it to the English agriculturist. The British climate is especially suitable for the growth of high class qualities of barley. This alone places the barley grower at a great advantage over the wheat grower, because the English climate is not suitable for wheat. Wheat in this country is always somewhat of the nature of an exotic plant, and wheat growers are always confronted with foreign wheats of a higher quality; but that is not the case with foreign barleys. Foreign barleys may be used, but you may depend upon it maltsters will prefer an English-grown barley if they can possibly get it. The somewhat humid character of this country is exceedingly suitable for growing a mellow sample of malting barley. These are important facts in contrasting the prospects of light land with the prospects of heavy land, and they offer good reasons why we should hope that cultivators of light lands will still be able to hold their own, if we can get such a correspondence between the expenses and the produce as to give us a balance upon the right side.

Another factor in the question is that light lands are not

naturally so well adapted for pasturage as heavy lands. They readily burn in hot weather, and the grasses are disposed to die out after a short series of years. The light land farmer is therefore less likely to take refuge in laying his acres down to permanent pasture.

Next with reference to the important subject of laying land down to grass, that again is beset with many difficulties. It is by no means an easy thing to produce that particular character of herbage which belongs to an old pasture. Everyone who is accustomed to country life knows—any one can tell who has any experience at all in farming—when he is walking over what is called a newly-laid down field. Even if it has been laid down a dozen or twenty years, I take it that we shall easily be able to judge that the ground is not yet in the beautiful rich condition that belongs to permanent pasturage. To secure a thoroughly good turf, in spite of what has been said of late years, I believe it requires time. I know that we have been told recently that all depends upon getting a proper mixture of grass seeds. Mr. Faunce de Laune has come forward conspicuously as an apostle of the importance of pure seeds. The Royal Agricultural Society has also taken up the subject, and we have a society's botanist in order to direct the members in the selection of pure seed. It is of vast importance not only that pure seed should be used, but that farmers should be well instructed as to the species of plants which ought to be sown in order to make a good pasture. But agriculture is a most extraordinary business. It is different to anything else, and we still find men, whose opinions we are bound to respect, express views contrary to what our scientific leaders tell us. It is extraordinary how often we hear that land which has not been laid away, but which has what is called "fallen away" itself, proves the best pasture in the end. We shall also find, as I was informed by a very eminent agriculturist the other day, that grass land, laid away with all the care and all the science he could muster, was yet inferior to grass land sown—with what?—with the hay-seeds of the district got up out of the hay-loft. I would not like to teach that such a system is the right one for laying land down to grass. It is unscientific. But when we purchase grass seed from an eminent firm we must remember that they may be pure, and they may be the right species, but they are not acclimatized to our district, and there are such numberless small and imperceptible differences in vegetation that there may be just that sufficient difference, owing to climatic causes, or owing to soil, why these expensive seeds disappoint us. Such a result is just possible; whereas we go to the hay-loft, and pick up the seeds which have been shot out of good hay —mind, really good hay, and that seed is grown in our own district; it contains a very large number of species, and it contains the species which the law of the survival of the fittest has proved to be fitting to our particular soil. Hence we see that from a scientific point of view there may be something in the sowing of the hay-seeds of the district, or even from a hay-loft, in preference to the sowing of expensive seeds ordered from seedsmen, and which may possibly have been grown out of England altogether. Not that I defend the system of sowing hay-seeds, because I think that it has its disadvantages also, but I think we ought to hold our theories and our scientific knowledge not too tightly with reference to a practical point of this kind. We must be open to conviction, and not make ourselves appear foolish before practical men by forcing ideas down their throats which they know by long experience are not according to fact.

Time is a most important element in the proper making of a permanent pasture. We must not forget the old adage that "to make a pasture will break a man." The converse in the old adage is worthy of notice, because it depends on the first—"to break a pasture will make a man." Now what does this piece of wisdom point to? It points to the fact that a pasture is a complete store of wealth. The amount of nitrogenous matter and available plant food in a pasture is extraordinary. That is the reason that to break a pasture will make a man, and so it will if he has sufficient of it. An old pasture is in the position of the virgin soils of the new world, and if a man has a sufficiency of it, it will make him wealthy. But it requires time to accumulate wealth both with people and with soils. It takes a long time, sometimes too long, and during the whole of that period which gradually brings the soil into that particularly rich state which belongs to permanent pasture, we must consider that money is being laid out, that is to say, it is done at a sacrifice. The man who lays down a large amount of pasture must be content to postpone his profits for many years, and perhaps it may even be said to impoverish the father and enrich the son. Lime is said to enrich the father and impoverish the son. The laying down of permanent pasture would have an opposite tendency; the father may have to do it at considerable expense and never see the advantage of it; but the son, coming into the occupation of his old pasture land, will at length reap the benefit.

There are agricultural teachers who scoff at such ideas, and say they can produce a better pasture in a very short time than the old pasture. The real fact is, that in laying down land to grass we get a full maximum the first year, and we get a very good crop the second year, and then I am afraid in very many cases the pasture begins to go back for, it may be, several seasons. That is the general history of laying down a pasture. There can be no doubt about it—there may be exceptions to the rule—but in the vast majority of cases your pasture will begin to go back; then, after a period, it begins to recover itself, and when it has been some twelve or fifteen years down, if it is successful, it ought to be growing into the condition of good permanent pasture. To shorten this period —to bridge it over—is our task; and by proper preparation of the ground, by employing the best methods of sowing, by selecting the best seed, and by judicious after treatment, we shall accomplish this object. I mention these points in order that we may discuss them *seriatim*.

With regard to the first point, it is no use sowing grass seeds upon worn-out poor soil, nor yet upon foul, filthy land. The ground should be well cleaned, and should be in a high state of fertility. In order to produce such a condition a root crop fed on the ground is desirable. As, however, we are supposing the case of clay land, a bare fallow may be substituted. Some persons have gone so far as to bare fallow land intended for permanent pasture two years in succession—a plan open to objection as an experiment likely to waste nitrogen. Still, the object is laudable, and the result may easily justify the means. The late Mr. J. Howard of Bedford adopted this system, and I have seen the excellent results he achieved on clay ground with the requisite liming and dunging and proper aërating cultivation which two years bare fallowing entailed. The object, whether that object is attained or not, is to bring the land into a thoroughly fertile state. It gives it two years' rest, two years' tillage, and good manuring. The grass seeds are sown in the month of August without a crop, that is, without an over-shadowing wheat or corn crop, as is customary. A root crop, or two successive root crops, would be more in accordance with the most modern notions, but would be inapplicable to the stiffest classes of soils. A root crop fed, or even two root crops in succession fed, would produce a state of things which would be highly conducive to a rapid formation of grass roots. I am quite convinced that the secret of success in laying down land to grass is in securing richness of soil, and this we may assure ourselves of by a little observation. Notice what naturally happens in the case of heaps made from pond-cleanings or ditch-bottoms recently cleared out. Here we find a quantity of rich stuff, and we may notice how rapidly such heaps become covered with a thick rich sole of grass. I have seen it very frequently. Here we have the best conditions for growing grass; we have richness, and without even sowing seeds it is extraordinary how those heaps of dirt have become covered with everything that we could wish for in the form of grassy sward. But the amount of richness, or of available plant food, and of nitrogen, is so great that it would be impossible to add it in similar quantities to large tracts of land. It would be altogether beyond us to attempt to confer that same amount of richness on fifty, or one hundred acres of ordinary arable land. All we can do is by bringing the ground into as rich a state as possible to encourage the matting of the grass and the formation of a close sole.

The next point is autumn sowing, and sowing without a crop. Sowing the seed upon land rich from previous management; sowing without an exhausting crop accompanying the grass seeds; and sowing in the early half of August. These conditions are likely to insure what we require. We must not be alarmed if weeds come up in such abundance with the grass seeds, that we are almost tempted to think our grass seeds are going to be smothered with chickweed, ground ivy, charlock, and other sorts of vegetable rubbish. I have seen it often, and have been even inclined to plough up the ground again; but it is wonderful how the grass will gradually assert itself, and these annual weeds—they are nothing more—die away and disappear. These arable land weeds require a finer tilth than they now have at their disposal, and the consequence is that the entire generation of annual plants dies away, and the grasses take their place.

With reference to the after treatment of pastures, perhaps the best after treatment that I can suggest is to let them grow without check in the spring of the year. Do not stock them; because as the stem develops upwards the root develops downwards, and the young plants spread without check and cover the whole surface. Many people cut the grass and make hay the first season, but the better plan is to turn cattle in when the crop is fit to mow. I believe we shall obtain a pasture if we do this. Turn the bullocks in, and let them trample it down or eat it—let them for once be knee-deep in grass. This may appear somewhat wasteful, and be rather expensive, but we shall get a pasture, and there is no great waste, as we can eat it down bare by and by. If the bullocks do not finish it, the horses and sheep will eat it down close before winter. Ground managed in this way is not unlikely to produce in a short time an excellent permanent pasture.

CHAPTER XV.

Proper Standard for Valuing Tillage Operations—The Cost of Maintaining a Farm Horse—The Cost of Tillages—The Cost of Growing a Crop of Wheat after Clover—Small Margin of Profit from Wheat Growing.

I HAVE hitherto treated in great measure of the soil, its origin, its composition, its varieties, and the methods by which it may be improved both in texture and in composition. We next passed on to the consideration of ordinary tillage operations, and afterwards, to the rotation of crops; and lastly, we considered the important subject of laying land down in permanent pasturage. I have in this last chapter to give a little attention to the subject of cost and realization. These are very important subjects in connection with agriculture and especially so at the present day, when we are frequently treated with estimates of cost in the daily press, in order to show the great difficulties under which English agriculturists are labouring. No doubt we stand at a very considerable disadvantage with regard to cost of production when we contrast our expenses with those incurred by the Indian cultivator, or the non-rent-paying American farmer. At the same time, there is reason for thinking that the estimates of the cost of producing a wheat crop in this country are often exaggerated. They are too often based upon a set of figures which have been adopted by agents or valuers, with the object of adjusting claims between out-going and in-coming tenants. But the theory upon which these figures rest is that the out-going tenant is allowed a profit upon every tillage operation; whereas in absolutely calculating the cost of a wheat crop, or of any other crop, the proper method of approaching the subject is to really inquire the cost that the various tillages entail upon the cultivator directly. If valuers' prices are taken, the cultivator is credited with a separate profit on each tillage operation, and can hardly expect a further profit upon the crop. The question is—What does it cost the farmer to plough an acre of land? not what will the valuer give the out-going tenant for having ploughed an acre of land for his successor? The difference between the two estimates is very considerable, and the amount which is usually charged in such cases appears to require revision.

Those estimates are based upon a higher scale of prices than now prevails, and they have been reduced very little, if at all. The amount which is charged for tillages by land-valuers at the present day is very much the same as was allowed for tillages in days of high prices—which scarcely appears reasonable. The method which I would suggest for arriving at the cost of tillages, would be in the first place to come to a sound conclusion as to the cost of horse labour. That is the key to the situation, because manual labour is an amount which can be very readily assessed; whereas the cost of horse labour is in itself a matter which can only be approximately arrived at. The cost of keeping a farm-horse lies at the very foundation of all costs of tillage; and without going too minutely into the subject, allowing market value for oats and hay, the cost of maintaining the farm-horse ought certainly not to be more than 9s. per week. Our estimate may be brought down to a definite statement as follows—

Cost of Keeping a Farm-Horse.

Winter feeding, per week.

	s.	d.
2 bushels of oats, at 2s. 3d. per bushel	4	6
1 peck (¼ bushel) of beans, at 4s.	1	0
1 stone (14 lbs.) of hay per diem, at £3 10s. per ton (that is, hay produced on the farm)	3	0¾
Straw, chaff, and litter	0	0
Total	8	6¾

Now with reference to summer keep. Summer keep is usually cheaper than winter keep, many farmers not giving corn during the summer months.

Summer keep, per week.

	s.	d.
Say, 1 bushel of oats	2	3
Cut clover, at 10s. per ton—say 56 lbs. per day ...	1	9
Grazing at night	2	0
Total	6	0

Thus we have brought out our estimate somewhat below what was originally stated. Now if we take the winter as extending over eight months or thirty-two weeks, we have—

	£	s.	d.
32 weeks at 8s. 6d. ...	13	12	0
20 ,, ,, 6s. ...	6	0	0
Total	£19	12	0

or say £20 for feed.

I have seen many estimates of food taken out in more detailed manner, but I have not seen any detailed estimate coming up to more than £24, and this after examining the literature of the subject for the last twenty-four years. Take, for example, the late Mr. Morton's most excellent book, *The Farmer's Calendar*, where he produces a large number of calculations as to the cost of keeping a farm-horse in food, and we shall not find many people place it higher than £24. Considering that we have experienced such a great reduction in the price of corn, I am not surprised that we should find £19 12s. to be the price required. But you must remember that there is the manure left. That is to say, we take the oats at 2s. 3d. or the beans at 4s. a bushel, when consumed upon the ground. That is the reason why I ask you to consider the straw, chaff, and litter as *nil*. It is not that we charge nothing for them, but we put the dung that is left against them. We are therefore fairly clear in starting with the cost of food at about £20. Then you have to add the following items of expense—

	£	s.	d.
Food, say	20	0	0
Shoeing, &c., put liberally at	1	0	0
Harness	1	10	0
Interest 5 per cent. on £30	1	10	0
Risk and depreciation at 10 per cent. on £30 ...	3	0	0
Total (including veterinary attendance)	£27	0	0

I do not know of any other item of cost to bring against the horse. Therefore we have nearly approached the figure which I originally started with. When I was first associated with farming some thirty years ago, we considered it cost £30 to keep a farm-horse. Since that time, during the high prices and during more prosperous times, I have heard good farmers endeavour to put the cost a great deal higher; but I could, I think, challenge any one to show me that a farm-horse, taken apart from wages, could possibly cost more than £30.

Now with reference to the question of labour. I know that certain estimates as to the cost of the farm-horse include the labourer, and a calculation occurs to my mind, namely, that of the late Mr. John Algernon Clarke, who wrote very ably on steam cultivation, and what can be saved by putting down a farm-horse. He calculated that you saved £44 a year by dispensing with a farm-horse in favour of steam—

that of course included the share of the ploughman's wages which goes with each horse.

I take then £30 as being a covering estimate of the cost of maintaining a farm-horse. Next we ask this question—How many working-days are there in the year? A great deal depends upon that. We must take out Sundays, and the two or three holidays the poor labourers enjoy, which are very few; they get Christmas Day and Good Friday—I am not sure all get that—and they get perhaps a club-day, but they have very few holidays. Sundays come out, then half and quarter days during which work is interrupted by wet, to set against which there are some long late days in hay-time and harvest. The most serious loss results from the fact that during frost and snow, and during continued wet weather, the horse is compulsorily idle, and that for quarter and half days as well as whole days.

If we take out fifty-two Sundays and, say, four holidays, and for broken days say ten, that leaves us 299 or 300 days, a figure which simplifies the matter very much if we may consider it as correct. I submit this estimate to your own judgment, as you may modify the calculation if you think that it is not according to experience. Starting with 300 days in every year, I am afraid we shall bring the cost per day down to a rather low figure. £30 equals 600s., and 300 days × 2 = 600; so that our result comes out 2s. a day. I believe that to be reasonable—that it costs the farmer about 2s. a day to keep his horses. It is not many ordinary farmers that give their horses such feeding as that proposed for our horses. How many give them no corn at all in summer? And in the winter they have to put up with straw, two bushels of oats, and a quarter bushel of beans. So that I think we are beyond the cost rather than within when we say 2s. a day. When I was a student under the late Professor Coleman he fixed the price for our purpose as students at 2s. 6d. a day. Then, I remember in days of greater prosperity some people were inclined to put it at 3s. a day. That is too high, and 2s. 6d. a day is too high; but at the same time many persons would assume that figure, and I do not see there is any great harm in assuming 2s. 6d. except that it is too high.

Next we proceed as follows. Ploughing is a very typical agricultural operation, and the unit by which you should measure ploughing is this—

2 horses and 1 man or lad will plough $1\frac{1}{4}$ acres of light loose land in 1 day. I have known more. I have known a Yorkshire lad plough 2 acres in a day. American farmers look for 2 and $2\frac{1}{4}$ acres to be ploughed in a day.

2 horses and 1 man or lad will plough 1 acre in 1 day of average land.

3 horses, 1 man, and 1 boy will plough from $\frac{3}{4}$ to 1 acre of stubble in a day.

4 horses, 1 man, 1 lad will plough $\frac{1}{2}$ to $\frac{3}{4}$ acre in a day of the stiffest land.

The cost, therefore, of ploughing depends on the stiffness of the land. It is not necessary, and I do not think it is advisable, for me at the present time to take each of these as an arithmetical sum. I leave that to you, or you can vary according to the requirements; but we will take the average —two horses and a man. Taking the horses at 2s. each, there is 4s.; take the wear and tear of the plough at 1s. which is quite sufficient, and take the wages of the man at 2s., that makes 7s. I consider that the average cost of ploughing when you can do an acre a day is 7s. an acre, and you will find that the price will rise with the stiffness of the ground and with the number of horses which may be required to draw the plough. If then you can get labour—and I think you can—at 2s. a day, you have 7s. an acre as the standard cost of ploughing.

We apply the same system of valuation to all tillages, as, for example, harrowing. A team ought to harrow from twelve to sixteen acres in a day. Now taking it at twelve acres a day, and putting a lad and a pair of horses to the harrow. We have 4s. for the two horses, 1s. for the lad, and 1s. for the use of the implement for the day, that is 6s. That is why it is that harrowing, we say, costs 6d. per acre.

In the same way with reference to rolling. A team will roll from twelve to thirteen acres per day. If it is a three-horse roller you have 6s. for the horses and, say, 2s. for the man—that is 8s. That is about 8d., or if a driver is employed, 9d. per acre. That is the principle on which we value tillages, and I thought it necessary to clear up this matter before taking out the cost of such an important item as that of producing a wheat crop. Now let us try to come to a conclusion on this matter. The cost of producing an acre of wheat after *clover*.

The first thing to decide is as to whether the clover crop has paid its own way, and we may assume that it has. The clover crop ought to pay its own way. In the last Agricultural Commission there are a great many estimates of the cost of producing crops, and it is very strange to see the unanimity with which the clover crop, at all events, is considered to be a paying crop in itself. The cultivation is light, and 1½ tons of hay with grazing is a very good return. The clover crop therefore is out of debt, and more than that, it has left a profit.

That being the case, it is not necessary to charge the wheat with any previous cultivation. At all events, we will not do so, as we might have done, and as we would have done if the wheat were taken after turnips. First, then, there is dung-carting, including filling, carting, and spreading; that we shall put at 10s., although a good deal depends on the distance to which we have to cart the dung.

ENGLISH AGRICULTURE.

Cost.

	s.	d.
Filling, cartage, and spreading of manure	10	0
Ploughing with 3 horses and a driver, 1 acre per day	10	0
Pressing, one-third the cost of ploughing	3	4
Two harrowings with heavy drags, 4 horses, at 1s.	2	0
6 harrowings at 6d.	3	0
1 drill (4 horses)	1	10
1 harrowing after drilling		6
Seed, 2½ bushels at 4s. 6d.	11	3
Bird-scaring, which ends autumn cultivation	1	0
Spring harrowing, say	1	0
Spring rolling	1	0
1 cwt. nitrate of soda at 11s. } 2 cwt. superphos. at 3s. 6d. }	18	0
Harvesting	15	0
Threshing	8	0
Dressing	1	0
Marketing	5	0
Rent, rates, and taxes, say	1 15	0
Total £6	6	11

Now as to the possible returns. There appears to be room for a profit even now, and in fact there must be if wheat cultivation is to continue. What can reasonably be expected from land costing 35s. an acre for rent, rates, and taxes, in wheat after clover, with a dressing of 1 cwt. of nitrate of soda and 2 cwt. of superphosphates to the acre, besides a dressing of dung? What should you look for? I think from nine to ten sacks would be the probable crop—thirty-six to forty bushels of corn, the straw to be considered as set against the cost of the dung.

	£	s.	d.
36 bushels of wheat at 4s. 3d. per bushel =	7	13	0
or if we take 40 bushels	8	10	0

This result shows that the margin to be looked for at present prices is small indeed. The fact is, that wheat cannot be grown at a profit at 17s. per sack. But if we value it at £1 per sack, or 5s. a bushel, the complexion of the calculation is a good deal altered for the better. A few years ago we used to receive 30s. a sack, or 7s. 6d. a bushel; and early in the century prices went up to 12s. or 15s. a bushel, and the expenses were much the same as now with the exception of the seed. Let us then hope for better prices and better times for English agriculturists.

INDEX.

ACTION of natural forces on land, 114—119
Agricultural experiments, 179, 180
Agricultural teacher, proper function of, 19
Agriculture as a definite subject, 1
—— faulty teaching of, 18
—— relations of science to, 3
Autumn cultivation, 127, 128

BARE-fallowing, cost of, 122
Basic cinder, 176

CATCH crops, 139
Clay-burning, 109
—— method and effect of, 110
Claying, 111
Clay land, crops for, 190
—— not adapted for sheep, 201
—— present and past returns on, 200
Clays, presence of iron in, 30
Clover and rye grass, 197
—— mown and overfed, 192
—— peas substituted for, 206
—— sickness, 197, 204
"Condition" in soils, 47
Constituents of clay soils, 30
Corn cultivation, 136
Cost of bare fallowing, 122

Crops, catch, 138—140
—— for heavy land, 187
—— seed bed for, 188
—— transplanting, 189
Cultivation of land, 114
—— aeration by, 115
—— aided by natural forces, 116—120
—— autumn, 118, 128, 129
—— by application of dung, 118
—— by frost, 116
—— by weather, 119
—— corn, 136
—— effect on weeds, 115
—— for seed bed, 115
—— of light soils, 124
—— of "roots" on heavy land, 121—123
—— of "roots" on light land, 122

DEEPENED tillage, by trenching, 108
—— advantages of, 109
—— assists drainage, 111
—— destroys "pans," 111
—— gradual production of, 109
—— objections to, 108
—— subsoils, 108, 111
—— suitable in various cases, 111
Diseases of root crops, 196

Q

Drainage, 91
— advantages of, 94—96
— a means of getting water into land, 100
— by Elkington's system, 104
— by open ditches, 104
— causes of wetness, 93
— capillarity in, 103
— effects of, on crops, 107
— effects on health of farm stock, 108
— effects on health of population, 108
— effects on health of tillage, 108
— effects on heavy and light soils, 98
— evaporation of moisture, 99
— Government grants for, 108
— naturally dry soils, 92
— of springs, 105
— pulverization by, 97
— reciprocal action in, 100
— Smith's system of, 101
— the furrow system of, 101
— water table or "reservoir," 92, 103
— wetness in soils, 93

EXPERIMENTS, agricultural mode of conducting, 179
— at Rothamsted, 166—176, 207
— at Woburn, 207

FALLOWING land, 185
Farm horses, cost of, 218, 219
Fertility, condition of, 40—42
— indications of, 43—50

Fertilizers, farm-yard manure, 146—154
— effect of climate on, 158
— general manure, 148
— "law of minimum," 165
— nitrogeneous, 155
— phosphatic, 155
— special, 164
— varieties of, 161
Fodder crops, 132, 138

GENERAL manures, 148
Geological classes of soils, 89
— peaty, 90
— sedentary, 89
— transported, 89
— volcanic, 90
— order of Bagshot sands, 59
— oolites, 74—76
— Cambrian, 86
— chalk formation, 67, 70
— coal-fields, 83
— Fen country, 63
— greensand, 72
— lias, 78
— Laurentian, 86
— Silurian, 86
— London clay, 58, 65, 66
— magnesian limestone, 82
— marsh land, 63
— mountain limestone, 84
— new red sandstone, 60, 79, 80, 81
— old red sandstone, 86
— peat, 60
— recently formed soils, 61
— river beds, 61
— rocks, 55, 57
— weald, 73
— wolds, Lincolnshire, 62

INDEX.

Geological, wolds, Yorkshire, 67
—— position as an indication of fertility, 55
Grain crops, 132
Grass and clover crops, 132
Grass cultivation on heavy lands, 124

IMPROVEMENT of soils, 91—128
Indications of fertility, 43—50

LIME, functions of, 34
Lime in nitrification, 34
Limestones and marbles, 34
Limits of agriculture as a study, 3

MANURES, farm-yard, 146
—— general, 148
—— nitrogeneous, 155
—— phosphatic, 155
—— special, 163
Marbles and limestones, 34
Marls, their composition, 30
—— their properties, 30

PERMANENT pasture, expense of making, 212
—— from high-class seed, 211
—— increase of, 208
—— land fallen away to, 211
—— laying down land to, 209—213
—— Mr. Faunce de Laune on, 210
—— Royal Agricultural Society on, 210
—— seeded with "hay seeds," 211
—— time required to make a, 212

ROOT crops, 132, 133, 134
—— manures for, 135
—— various, 198
Rotation, a practical and scientific, 196
—— a six course, 193
—— East Lothian, 199
—— Norfolk, 133, 186, 193
—— modifications of Norfolk, 195
—— Northumberland, 196
—— position of potatoes in, 199
—— sainfoin, clovers in, 204
—— suitable for heavy land, 190, 201
—— three field course, 183, 193
—— the Holderness, 189
—— theory of, 203
—— cleaning land under, 206
—— physiologically considered, 206
—— maintains balance of work, 207
—— maintains balance of live stock, 207
—— purifies land for sheep, 208
Rothamsted, results of experiments at, 166
—— common salt on mangel, 173
—— farm-yard manure on mangel, 173
—— unmanured plots, 166
—— leguminous crops at, 176
—— mineral manures on mangel, 173
—— mixed mineral manures at, 168, 171
—— nitrate of soda on mangel, 172

INDEX.

Rothamsted, permanent pasture, experiments at, 174
—— superphosphates at, 166

SAND, micaceous, 32
—— mechanical effect of, 32
—— silicious or quartzoze, 31
Soils, as a laboratory, 39
—— chemical properties of, 26
—— "condition" in, 47
—— conditions of fertility in, 40—42
—— constituents of clay, 30
—— fertile, 29
—— geological position of, 54
—— gradation of, 33
—— indications of fertility, 43—50
—— insoluble part, 29
—— origin of, 20, 27
—— proximate constituents of, 28
—— soluble part of, 28
—— structure of, 24
—— use of stones in, 37
Stones, their use in soils, 37
Subsoiling and trench ploughing, 108—111

Subsoils, chalk, 53
—— effect of cropping on, 51
—— indications of fertility in, 52
—— influence of drainage on, 52
—— rocky, 53
—— top dressing, 53
Syllabus of crop cultivation, 140

TEACHER of Agriculture, functions of, 19
Tillages, standard for calculating cost of, 216, 217
—— cost of, 221, 222
Trifolium incarnatum, 138

VALUE of land, 88
—— of oat straw, 196
Vegetable matter, a source of carbonic acid in soils, 36
—— effect of, in texture, 37
—— in pastures, 35
—— in soils, 35
—— oxidation of, 36

WARPING, 113
Wheat after clover, 196; cost of, 222; profit of, 223
—— after roots, 196

THE END.

www.ingramcontent.com/pod-product-compliance
Lightning Source LLC
Chambersburg PA
CBHW022009220426
43663CB00007B/1017